MIOCENE FLORAS FROM THE MIDDLEGATE BASIN, WEST-CENTRAL NEVADA

Miocene Floras from the Middlegate Basin, West-Central Nevada

by Daniel I. Axelrod

UNIVERSITY OF CALIFORNIA PRESS
Berkeley · Los Angeles · London

UNIVERSITY OF CALIFORNIA PUBLICATIONS IN GEOLOGICAL SCIENCES

Volume 129
Issue Date: January 1985

University of California Press
Berkeley and Los Angeles
California

University of California Press, Ltd.
London, England

ISBN 0-520-09695-9
LIBRARY OF CONGRESS CATALOG CARD NUMBER: 84-2414

Library of Congress Cataloging in Publication Data
Axelrod, Daniel I.
Miocene Floras from the Middlegate Basin, West-Central Nevada.
(University of California publications in geological sciences;
v. 129)
Bibliography: p.
1. Paleobotany-Miocene. 2. Paleocology-Nevada-Middlegate Gap
Region. I. Title. II. Series
QE929.A88 1984 560'.1'78 84-2414
ISBN 0-520-09695-9 (pbk.)

Contents

Illustrations

FIGURES

PLATES

MAP

Tables

Acknowledgments

This work has been carried out with National Science Foundation Grant DEB 79-05843, which I gratefully acknowledge.

My thanks are extended to Harold Bonham, Nevada Bureau of Mines, for assistance in selecting volcanic rocks suitable for radiometric dating and for discussions of geologic problems in the field. The radiometric data were also used by Dr. Katherine J. Barrows in her study of the geology of the southern Desatoya Mountains and Eastgate Hills (Barrows, 1971). Students in my classes in paleobotany at the University of California, Davis, aided in collecting fossil plants at the Middlegate and Eastgate sites. Study of these dissimilar floras introduced them to the field of paleoecology and to the history of the modern ecosystems derived from these and other Miocene floras in the nearby region.

Finally, I want to thank Peggy Wheat of Fallon, Nevada, who some 20 years ago advised me of the occurrence of the Eastgate flora.

To Marjorie Hughes, I assign a gold star for her assiduous editorial work that brought the manuscript into its present shape. And special acknowledgment is due Lyn Noah who, with good humor, has suffered through the manuscript revisions and put this publication into its present form.

Abstract

This report describes two contemporaneous Middle Miocene (18.5 m.y.) floras that occupied Middlegate basin, Churchill County, Nevada. Situated only 8 km apart on opposite shores of a lake, they differ considerably in composition. The Middlegate flora of 64 species is dominated by sclerophyllous trees and shrubs while forest taxa are rare, since they lived in the bordering hills that faced south. The Eastgate flora of 55 taxa has a larger representation of forest species but sclerophylls are not so well represented, probably because the site of accumulation was near north-facing canyons that enabled forest taxa to descend close to the shore. Precipitation varied from 76-89 cm (30-35 in.) in the northwest to 89-101 cm (35-40 in.) in the southeast part of the basin, with 5-8 cm (2-3 in.) distributed each summer month. Climate was mild: mean July and January temperatures were approximately 20° and 5°C, respectively. There was light snow in the winter, with 4-5% of annual hours subject to freezing. The basin is estimated to have stood at near 915 m (3,000 ft.), as compared with 1,525 m (5,000 ft.) today.

Generally, contemporaneous as well as younger (15 m.y.) floras to the north, west, and south have more numerous exotic taxa. This seems related to local terrain that favored the ingress of moisture-laden air masses, whereas the Middlegate basin was in the rainshadow of hills that have since been largely removed by basin formation. Somewhat younger (13-12 m.y.) floras throughout the Great Basin, Columbia Plateau, and California are greatly impoverished as compared with those 15 m.y. and older; the younger floras have fewer exotics and are of subhumid aspect. This is attributed to the climatic consequences of the spread of the East Antarctic ice sheet, a phenomenon possibly resulting from the effects of episodic volcanism on global climate. Lowered sea-surface temperature would result in decreased summer rain in middle latitudes on the west coasts, and hence to reduced numbers of exotic taxa.

INTRODUCTION

The composition and distribution of vegetation in regions of temperate climate and moderate relief is determined primarily by local climatic conditions governed chiefly by slope relations and elevation. In such areas, the vegetation on opposite shores of a lake often shows marked differences in composition, as between forest types, or in physiognomy, with woodland replacing forest in drier areas. Changes of this nature occur today throughout the western United States and in many other regions as well. In view of these relations, we may anticipate that two contemporaneous fossil floras situated on the opposite shores of an ancient lake might also reveal important differences in composition, if the area was one of moderate relief and the climate was subhumid.

In this connection, two Miocene floras from the Middlegate basin in west-central Nevada assume considerable significance (fig. 1). The Middlegate flora, described earlier from a site on the northwest margin of Miocene Middlegate Lake (Axelrod, 1956), differs appreciably from the Eastgate flora situated only 8 km (5 mi.) southeast, on the opposite shore of the same lake. The Middlegate flora is dominated by sclerophyll forest taxa (*Arbutus*, *Chrysolepis*, *Lithocarpus*, *Quercus*), and conifer-hardwood forest taxa have a low representation. At the Eastgate site, conifer forest taxa (*Abies*, *Larix*, *Picea*, *Pinus*, *Pseudotsuga*, *Sequoiadendron*) are more abundant, and sclerophyll forest taxa are reduced in both

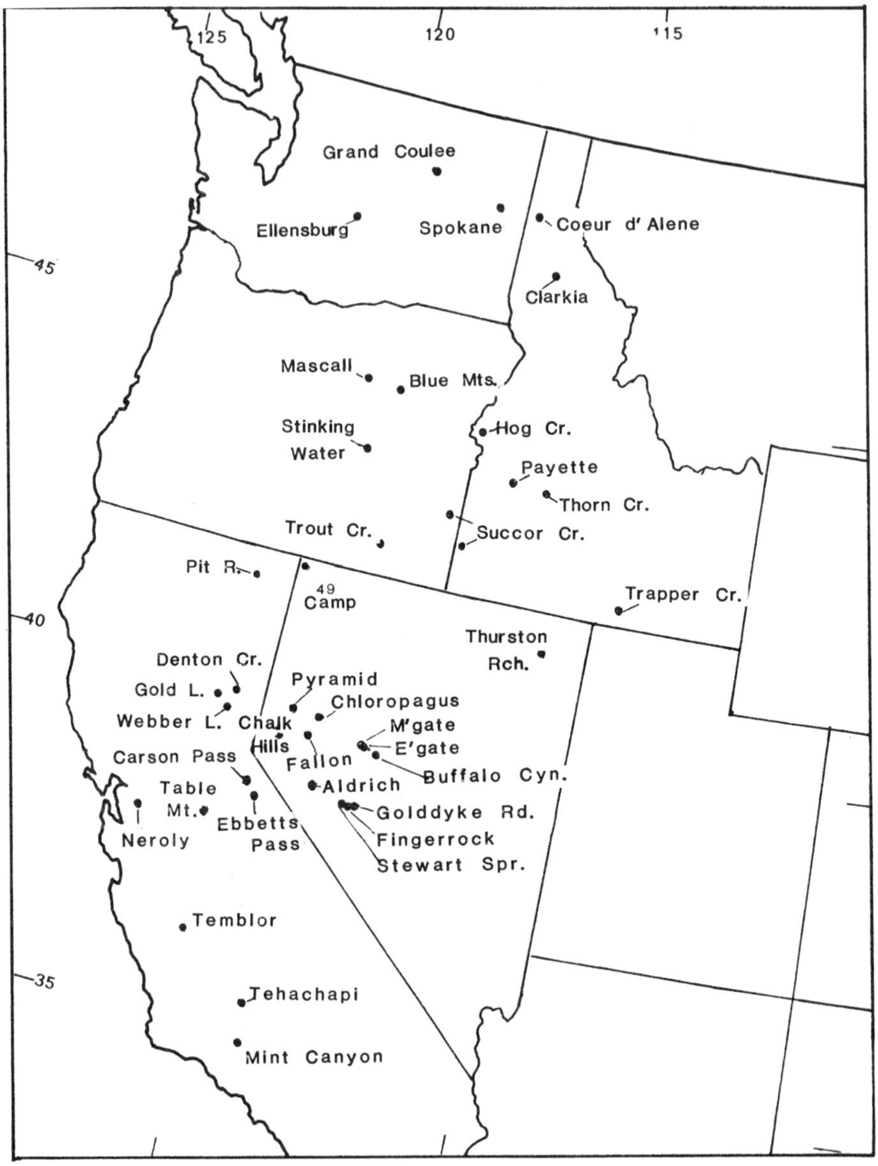

FIGURE 1. Location of Middlegate and Eastgate floras and some of the others referred to in the text.

diversity and abundance. Inasmuch as both floras are stratigraphically near the top of the Middlegate Formation, and hence of the same age, the differences between them seem attributable to local climates governed chiefly by terrain and exposure on opposite shores of the lake. Since these floras provide us with the first clear example of topographic-climate control on Miocene forest distribution in a very local area in the western United States, it is desirable to analyze them in detail.

The Middlegate flora is reviewed first; supplementary collections of 3,000-odd specimens have added 22 species to the 42 previously reported from it. This is followed by description of the composition and vegetation of the Eastgate flora. The data for both floras are then used to estimate the climate of the area and the altitude of Middlegate Lake. The provincial relations of the floras are then discussed.

Present Physical Setting

The two floras come from a small structural basin situated between the south end of the Clan Alpine Range and the Eastgate Hills, a subsidiary block of the high Desatoya Range. These are the first of a series of high fault-block ranges that dominate the landscape of central Nevada to the east, ranges whose summits rise to 3,000-3,600 m (10,000-12,000 ft.). To the west, the mountains are mostly lower, more irregular in outline, and separated by broad alluviated basins. Drainage in the area is west from the Desatoya Mountains, across Buffalo Basin, through the narrow gorge that cuts across the Eastgate Hills at Eastgate, thence across the Middlegate basin. Drainage continues west through Middlegate Gap that incises the south tip of the Clan Alpine Range, thence through the gorge at Westgate about 13 km (8 mi.) farther west, and then north some 65 km (40 mi.) to the sump of Humboldt Salt Marsh in Dixie Valley.

The lower parts of Middlegate basin are covered with a shadscale desert flora, typified by species of *Atriplex*,

Ephedra, *Eurotia*, *Grayia*, *Lycium*, and *Tetradymia*. These low, dense, rigid, usually spinescent shrubs are scarcely 0.5 m high, for the most part. The only tall shrub in the lowlands is *Sarcobatus vermiculatus*, which is confined to the moist drainageway in the lower part of the basin. Precipitation in the shadscale desert near Middlegate is about 125 mm (5 in.) annually. It gradually increases to 200-225 mm (8-9 in.) on the lower windward slopes of the Eastgate Hills, where the upper parts of the alluvial fans are covered with pure stands of Basin sagebrush (*Artemisia tridentata*), a shrub about 1 m tall. Juniper (*Juniperus osteosperma*) and pinon (*Pinus monophylla*) appear locally at levels near 1,670 m (5,500 ft.). These are stragglers from the Eastgate Hills that reach down drainageways and on cooler northerly-facing slopes to the Eastgate fossil locality in the southeast part of the basin. This conifer woodland assumes dominance where precipitation totals about 350-450 mm (15-18 in.) annually.

To the east, across the narrow, linear Eastgate Hills, the broad Buffalo Basin lies at the front of the Desatoya Range. This basin stands at a general level of 1,680-1,800 m (5,600-6,000 ft.) and is covered with a pure *Artemisia* association. Stringers of juniper and pinon reach down into the basin, especially at the south, where the valley floor rises to higher levels. Conifer woodland covers most of the Desatoya Range except for its highest parts, generally above 2,580 m (8,500 ft.), where *Artemisia* dominates. The cooler, moister parts of the range support scattered stands of aspen (*Populus tremuloides*). Large shrubs, notably species of *Alnus*, *Prunus*, *Ribes*, *Rosa*, and *Salix*, choke the narrow canyons and form dense patches at seepages on the slopes.

Precipitation comes chiefly in winter as rain and snow, with January the wettest and coldest month. However, violent summer showers occur locally and infrequently in the mountains during summer. The resulting flash-floods may deeply erode the natural drainageways and remove desert highways and tracks.

As judged from an incomplete record (1957-66), the mean annual temperature at Eastgate is now about 10.8°C (51.5°F), with a mean range of approximately 23.3°C (42°F), from a July mean of 22.2°C (72°F) to a January mean of -1.1°C (30°F). Extreme high summer temperatures reach 39.4°C (103°F), and extreme winter lows fall to about -8°C (17°F). These data give the area a warmth (or effective temperature) of *W* 13.2°C (55.7°F)--or a growing season of 158 days with mean temperatures above that degree--and an equability (or temperateness) rating of *M* 47 (on a scale of 100).

Vegetation Terms

In order to discuss the fossil floras and to reconstruct the vegetation that they appear to represent, it is essential to use a consistent terminology. Unfortunately, there has been little consistency in the terminology on vegetation in the western United States. Physiognomic terms dominate, but these are notoriously inexact. Not only have divergent names been applied to woodlands and forests of the same general composition, some of the terms are misleading. For example, mixed-evergreen forest might be considered by some to refer to a forest rich in conifers (=evergreens), though it is also applied to broadleaved sclerophyll vegetation. To some, the term mixed-conifer forest designates a forest composed of several conifers (say *Abies*, *Pinus*, *Pseudotsuga*), yet to others it refers to a forest of conifers and deciduous hardwoods. To eliminate any ambiguity, I shall use these terms as defined below.

Sclerophyll Forest

This refers to a forest of relatively low stature dominated by broadleaved evergreen dicots, notably species of *Arbutus*, *Chrysolepis* (=*Castanopsis*), *Lithocarpus*, and *Quercus*. These are usually associated with evergreen shrubs, including species of *Arctostaphylos*, *Ceanothus*, *Cercocarpus*, *Heteromeles*, *Schmaltzia* (=*Rhus*), and others that may also contribute to shrub communities ("chaparral") on local dry sites. Associates of the sclerophyll forest may also include deciduous taxa, especially along streams and at seepages, such as *Acer*, *Fraxinus*, *Juglans*, *Platanus*, *Populus*, *Ribes*, *Rosa*, and *Salix*.

Conifer-Hardwood Forest

This denotes a fossil assemblage of tall forest conifers (*Abies*, *Larix*, *Picea*, *Pinus*, *Pseudotsuga*, *Sequoiadendron*, *Chamaecyparis*) and associated exotic conifers, either such genera as *Ginkgo*, *Glyptostrobus*, and *Taxodium*, or species of *Abies*, *Pinus*, *Picea*, *Larix*, and others. In addition, the forest includes exotic deciduous hardwoods, notably species of *Acer*, *Alnus*, *Betula*, *Carya*, *Hydrangea*, *Nyssa*, *Ostrya*, *Platanus*, *Ulmus*, and *Zelkova*. Associated shrubs are chiefly deciduous, such as *Amelanchier*, *Crataegus*, *Rhus*, *Ribes*, *Rosa*, and *Sorbus*, some of which may be exotic. Evergreens may be present, either as exotic trees (*Magnolia*, *Persea*) or shrubs (*Ilex*, *Mahonia*), or as surviving, derived natives (*Chrysolepis*, *Heteromeles*, *Mahonia*).

Mixed-Conifer Forest

This term denotes the diverse *modern* forests dominated by mixtures of western conifers noted above, as well as their usual associates that also survive today in the western United States, including species of *Alnus*, *Betula*, *Amelanchier*, *Chrysolepis*, *Fraxinus*, *Mahonia*, *Platanus*, *Populus*, *Ribes*, and *Vaccinium*, and many others. There was a gradual transition from conifer-hardwood to mixed-conifer forest as exotics were gradually eliminated during the Neogene, a shift that commenced first in the interior.

Summary of the Described Floras

The geographic occurrence, composition, vegetation, and age of the two floras described in this volume are summarized as follows.

Middlegate Flora

Situated 3.5 km northeast of Middlegate Gap, in the upper 10 m of the Middlegate Formation.

The sample consists of 6,882 specimens distributed among 64 species, of which 22 are new to this previously described flora.

The taxa contributed to a dominant sclerophyll forest, with conifer-hardwood forest taxa rarely represented.

Age is late Hemingfordian, dated by K/Ar at 18.5 ± 2.2 m.y. on hornblende (Geochron, Inc.).

Eastgate Flora

Situated in the badlands 5 km south-southwest of Eastgate, in the upper 10 m of the Middlegate Formation.

The sample consists of 5,805 specimens distributed among 55 species, of which 13 are new.

The taxa contributed to the lower part of a conifer-hardwood forest, with a restricted sclerophyll forest nearby.

Age is Hemingfordian, dated at 18.5 m.y., as above.

GEOLOGY

Noble (1972) related the episodes of volcanism in this region to changes in relative movements of the North American and Farallon plates and the underlying asthenosphere. Oligocene (33-29 m.y.) quartz latitic and rhyolitic ash-flow volcanism in the east-central and central Great Basin may represent the later stages of an early phase of calc-alkaline volcanic activity. In a second pulse of igneous activity during the early Miocene, rhyolitic ash-flow tuffs and subordinate lavas erupted along an arcuate belt concave to the northeast, extending from eastern Oregon to southwesternmost Utah. West and southwest of this axis of early Miocene silicic volcanism is a series of middle Miocene and younger lavas, mainly andesitic and dacitic with minor quartz latite pyroclastic rocks, that represent a southeastern extension of the western Cascade belt of subduction-related volcanism. The Middlegate area lies on the boundary of the early Miocene silicic and middle Miocene intermediate volcanism outlined by Noble (1972).

The Middlegate area is part of one of the two major volcano-tectonic troughs that typify the central Great Basin (Riehle, McKee, and Speed, 1972; Burke and McKee, 1979). It is bounded on the north and west by the low south end of the Clan Alpine Mountains, and on the east and south by the arcuate Eastgate Hills, a subsidiary block of the Desatoya Mountains which lie to the east. The ranges are composed of rhyolite, quartz latite, and dacite welded tuffs, breccias, and flows over 3,000 m thick (Barrows, 1971; Noble, 1972; Burke and McKee, 1979). In the area near Westgate, a few km west of Middlegate, the volcanic rocks rest on highly folded and metamorphosed Triassic and Jurassic marine clastics and carbonates, and are intruded by granitics. Similar relations appear in small structural windows at the northern and southern ends of the Desatoya Mountains (Willden and Speed, 1974).

Two sedimentary formations of Miocene age rest on the volcanic rocks of the Middlegate basin. The older Middlegate Formation contains the Middlegate and Eastgate floras in its upper part. It is overlain unconformably by the Monarch Mill Formation, which has a large mammalian fauna of early Barstovian age in its basal beds (fig. 2 and Geologic Map).

Middlegate Formation

The Middlegate Formation is exposed around the margins of the Middlegate basin, being covered elsewhere by the overlying Monarch Mill Formation or by Quaternary alluvial deposits. Its type area is on the west and north sides of the basin, where the Middlegate flora was recovered. There, a basal conglomerate is present discontinuously, its distribution corresponding to sites where streams entered the lake from nearby hills. Elsewhere the basal beds are sandstone, mudstone, or shale, and thin rhyolitic tuffs are present locally. Most of the formation on the west side of the basin

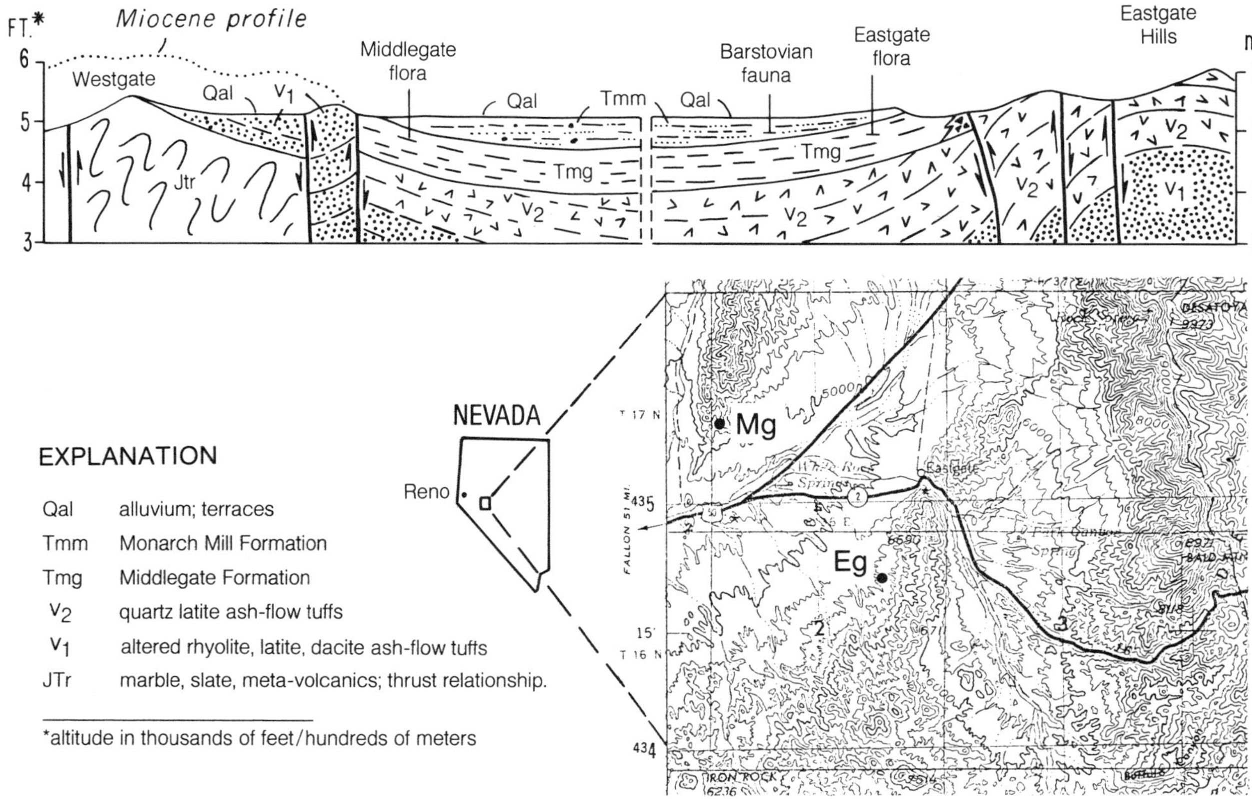

FIGURE 2. General topographic-geologic setting of the Middlegate and Eastgate fossil floras.

is composed of well-bedded, cream to white, semi-opaline shale separated by thin, light-yellow bentonitic clays that represent altered ashfall deposits. The formation is about 400 m (1,300 ft.) thick, and on the west side of the basin is overlain, probably unconformably, by the Monarch Mill Formation, composed of sandstone, shale, and some pebble conglomerate.

The Middlegate flora is preserved in opaline shales interbedded with bentonite in the upper 10 m (30 ft.) of the formation (Plate 1, fig. 1). From the locality, the section grades north into dark gray mudstones and thence into a cobble conglomerate that was laid down at the mouth of a valley in the volcanic basement hills. The fossil site was thus situated south of a valley that drained the south-facing slope of a low range. The concentration of plant structures at the site (thousands of leaves are present) evidently corresponds to the distance (about 0.8 km [0.5 mi.]) they were carried by currents in the lake. The lake was in the lee of a range to the west, the remnants of which are now the low hills north and south of Middlegate Gap. The Miocene hills have been sundered by faulting and now underlie the alluviated valley to the west, having been downdropped by faulting following deposition of the sedimentary formations.

There is a marked change in lithology and structure as one proceeds to the east side of the basin south of Eastgate. There the Middlegate Formation is buttressed against the older volcanic rocks along the Eastgate fault. The basal beds represent a thick sedimentary breccia with megaclasts commonly several feet in diameter. The clasts that make up the breccia vary in composition along the margin of the basin, reflecting the nature of the volcanic terrain in the bordering hills. Southward from Schweiz Canyon (sec. 11), the fanglomerate is monolithologic, being composed of blocks of welded quartz latite tuff derived from the hills behind the scarp. A mile to the north, the clasts include the dark-colored, older Eastgate Volcanics, composed of altered quartz latite flows and tuffs, and also some reworked conglomerate associated with the volcanic sequence. There can be little doubt that the breccia was derived from a rising scarp behind the Eastgate fault, which was active during deposition of the lower part of the formation. Major landslide blocks from 10 to 50 m long are surrounded by opaline shale in the southeast part of the basin (sec. 14). Renewed movement on the fault accounts for the absence of the breccia along the scarp directly east of the plant locality.

The overlying section is composed of alternating shale, thick sandstone, small-pebble conglomerate, and thin, fine-grained tuffs. Persistent peaty beds 1-2 m thick occur near the middle of the formation east of the Eastgate fossil locality. The Eastgate flora is preserved in buff to reddish-brown shale about 10 m (30 ft.) below the top of the formation (Plate 1, fig. 2). The overlying conglomerate represents the basal member of the Monarch Mill Formation in this area.

The pebble conglomerates in the Middlegate Formation at the Eastgate locality are composed chiefly of rhyolitic clasts derived from the Eastgate Hills to the south. Clasts that might have come from the Desatoya Range east of the Eastgate Hills have not been found in the Middlegate conglomerates or in those of the overlying Monarch Mill Formation. Among the conspicuous rocks in the Desatoya Range that are absent from the Middlegate Formation are black vitrophyric andesite of the Tortoise Peak Formation, crystal-poor reddish latite of the Skull Formation, and light-gray to white welded rhyolite tuff of the Carroll Formation, all described by Barrows (1971). Inasmuch as these rocks in the Desatoya Mountains differ in composition and appearance from those in the Eastgate Hills, and would easily be recognized if seen, it is evident that the Eastgate and Buffalo basins were separated by the elevated Eastgate Hills during deposition of the Middlegate Formation. This is consistent with the nature of the basal breccias of the Middlegate Formation, for they were derived from a rising scarp along the west front of the

Miocene Eastgate Hills. Furthermore, examination of current bedding in the channel sandstones of the Middlegate Formation just below the fossil locality shows that their source area lay to the south or southeast. The evidence thus indicates that the Middlegate basin was separated from Buffalo Basin, in which the Buffalo Canyon flora lived, by the narrow, steep Eastgate Hills.

Monarch Mill Formation

The Monarch Mill Formation, which overlies the Middlegate Formation, was described originally from exposures in the middle and western parts of the Middlegate basin (Axelrod, 1956, p. 186). It has since been traced into the east and southeast parts of the area, where its clastic sediments are much coarser. There the basal part of the formation is a widespread sedimentary breccia that grades basinward into a cobble or small boulder conglomerate 4-7 m thick. Conglomerate is also scattered throughout the formation, often associated with thick, poorly indurated, poorly bedded, coarse and pebbly sandstone in beds up to 9-10 m thick. Shale and mudstone beds are present, as are occasional fine-grained tuffs, but they are largely masked by rubble-covered slopes that are capped by unconsolidated alluvial fan deposits. The formation earlier yielded a few scrappy mammalian remains (chiefly limb bones) at a site in the middle of the basin (Axelrod, 1956, p. 204). More recently, a large mammalian fauna was recovered from the basal part of the formation, about 0.8 km south of the Eastgate flora; this is discussed in the chapter on age.

Physical Setting

The marked differences in sedimentary rocks and structural relations on the east and west sides of the Middlegate basin indicate that topography of the areas where the Middlegate and Eastgate floras lived differed appreciably. The Middlegate flora is associated with fine-grained sedimentary rocks, notably siliceous shale, fine-grained tuff, and mudstone. The siliceous shale evidently formed in clear, quiet water in the lee of low hills. Coarse clastics are rare in this section. As noted earlier, sandstone occurs locally in the basal part of the formation, and conglomerate is also present there. Their distribution indicates that they represent thin tongues restricted to sites where streams entered the lake from hills to the north. That the nearby terrain was low, and remained low, during deposition of the succeeding Monarch Mill Formation is implied by the fine-grained sedimentary rocks that lap onto the older volcanic terrain on the northwest side of the basin.

Both the Middlegate and Monarch Mill formations change toward the south and east parts of the basin. Replacement of the opaline shale of the Middlegate area by mudstone in the Eastgate region implies that the lake was more turbid and agitated in the latter area. This is expectable, inasmuch as the Eastgate flora was on the western, windward slopes of the ancestral Eastgate Hills. That the Eastgate fault was active during deposition of the Middlegate Formation is shown by the occurrence in its lower part of breccia and conglomerate which are buttressed against the fault. Another difference is that the overlying Monarch Mill Formation has coarse sedimentary breccia and cobble-boulder conglomerate, as well as thick sandstone beds, all absent near the Middlegate locality. Clearly, the east margin of the basin was bordered by steep hills that were tectonically active during deposition of both formations, whereas the relief was relatively low to the west and northwest at the south end of the ancestral Clan Alpine Range, where the Middlegate flora lived.

The low, rolling hills that formed the north and northwest margin of the Middlegate basin largely faced south, and hence were effectively dry and warm. Since hills bordering the Eastgate locality faced west, they provided that area with drier sites on the west- and southwest-facing slopes, and that these were relatively steep is implied by the coarser detritus in this section. Hence, some Eastgate sites were moister and cooler than others at similar distances from the site of the Middlegate flora. Clearly, the Eastgate area would be expected to support a flora of more mesic aspect. This relation was accentuated by the different position of the floras to the hills that rimmed the area. Since the Middlegate flora was in the lee of hills to the west composed of older volcanic rocks, it was in a local rainshadow with respect to moisture-bearing winds moving into the area. The Eastgate area would have received more precipitation because it was on the western, windward front of the ancestral Eastgate Hills. Some additional moisture probably would be picked up as air moved across the lake. Furthermore, its windward position would moderate temperatures, so that precipitation would be more effective there. Even today, the Eastgate area on the windward side of the Eastgate Hills has a more mesic flora than the Middlegate area. Whereas juniper and piñon reach well down into the basin on the western, windward slope of the central and southern Eastgate Hills, near Middlegate there is only a desert shrub flora at much higher levels on the lee slopes of the drier, southern Clan Alpine Range. During the Miocene, the difference in precipitation may have amounted to 125-250 mm (5-10 in.) in effective moisture.

To summarize, the Middlegate flora lived on south-facing slopes in the lee of the south end of volcanic hills ancestral to the Clan Alpine Range, a setting that gives a warm and dry aspect. By contrast, the Eastgate flora covered the steep, western, windward slopes of the north-trending Eastgate Range. There were some dry sites, but the cooler, north-facing slopes and deep sheltered canyons favored the presence near the lakeshore of mesic communities as well as those with cooler requirements from somewhat higher levels. The differences in topographic setting are reflected in the composition of these contemporaneous fossil floras that are situated only 8 km (5 mi.) apart.

THE MIDDLEGATE FLORA

Composition

Since this flora was first described (Axelrod, 1956), additional collecting with the aid of student classes on several occasions since 1968 has added 22 species to the assemblage, and field counts have nearly doubled the size of the previously published quantitative data. This adds importantly to our understanding of this flora, which is now known from 6,882 specimens. A revised list appears below, with new taxa marked by an asterisk (*). Several previously listed species are absent here, their status having been altered for diverse reasons presented in the systematic section and elsewhere (Axelrod, 1980).

As now known, the Middlegate flora is composed of 64 species. Apart from 3 aquatic or semi-aquatic taxa (Equisetum, Nymphaeites, Typha), the assemblage includes 10 conifers and 51 woody dicots, as compared with 5 conifers and 34 dicots reported earlier. Doubling the size of the sample has thus added 5 conifers and 17 dicots to the flora. Although many investigators would consider the initial sample of 3,400 specimens more than adequate, it clearly was not representative of the flora in the nearby area. A sample probably can be considered representataive if, after two full days of digging, no additional taxa are added to the collection.

Systematic List of Species

Equisetaceae
- Equisetum alexanderi Brown

Pinaceae
- Abies laticarpus MacGinitie
- *Abies scherri Axelrod
- Picea lahontense MacGinitie
- Picea sonomensis Axelrod
- *Pinus sturgisii Cockerell
- *Pseudotsuga sonomensis Dorf
- *Tsuga mertensioides Axelrod

Cupressaceae
- Chamaecyparis cordillerae Edwards and Schorn, new sp.
- *Juniperus nevadensis Axelrod

- cont. -

Taxodiaceae
 Sequoiadendron chaneyi Axelrod

Typhaceae
 Typha lesquereuxi Cockerell

Salicaceae
 *Populus bonhamii Axelrod, new species
 *Populus cedrusensis Wolfe
 Populus eotremuloides Knowlton
 Populus payettensis (Knowlton) Axelrod
 Populus pliotremuloides Axelrod
 *Salix owyheeana Chaney and Axelrod
 Salix pelviga Wolfe
 Salix storeyana Axelrod
 *Salix venosiuscula Smith
 Salix wildcatensis Axelrod

Juglandaceae
 Juglans nevadensis Berry

Betulaceae
 *Alnus harneyana Chaney and Axelrod
 *Alnus largei (Knowlton) Wolfe
 Betula thor Knowlton
 Betula vera Brown

Fagaceae
 *Chrysolepis convexa (Lesq.) Axelrod, new comb.
 *Chrysolepis sonomensis (Axelrod) Axelrod, new comb.
 *Lithocarpus nevadensis Axelrod, new species
 Quercus hannibali Dorf
 Quercus shrevoides Axelrod, new species
 Quercus simulata Knowlton

Berberidaceae
 Mahonia macginitiei Axelrod, new species
 *Mahonia reticulata (MacGinitie) Brown
 Mahonia simplex (Newberry) Arnold
 Mahonia trainii Arnold

Nymphaeaceae
 Nymphaeites nevadensis (Knowlton) Brown

Platanaceae
 Platanus dissecta Lesquereux
 Platanus paucidentata Dorf

Hydrangeaceae
 *Hydrangea ovatifolius Axelrod, new species

Rosaceae
 Cercocarpus antiquus Lesquereux
 Cercocarpus eastgatensis Axelrod, new species
 Crataegus middlegatensis Axelrod
 Crataegus pacifica Chaney
 *Heteromeles sonomensis (Axelrod) Axelrod, new comb.
 Lyonothamnus parvifolius (Axelrod) Wolfe
 Prunus moragensis Axelrod

- cont. -

*Sorbus idahoensis Axelrod, new species

Leguminosae
*Gymnocladus dayana (Knowlton) Chaney and Axelrod
Robinia californica Axelrod

Aceraceae
Acer middlegatensis Axelrod
Acer negundoides MacGinitie
*Acer nevadensis Axelrod, new species
Acer oregonianum Knowlton
Acer scottiae MacGinitie
Acer tyrrelli Smiley

Meliaceae
*Cedrela trainii Arnold

Rhamnaceae
Ceanothus precuneatus Axelrod

Stryacaceae
Styrax middlegatensis Axelrod

Ebenaceae
*Diospyros oregonianum (Lesq.) Chaney and Axelrod

Ericaceae
Arbutus prexalapensis Axelrod

Oleaceae
Fraxinus coulteri Dorf
Fraxinus millsiana Axelrod

---

*Species new to flora.

Judging from the usual habit of their nearest modern allies (Table 1), the flora includes 36 trees (10 conifers, 26 dicots), 25 shrubs or small trees (17 deciduous, 8 evergreen), and 3 aquatic to semi-aquatic herbaceous perennials. Although most of the Middlegate species are very similar to living plants, one of the woody dicots, _Lyonothamnus parvifolius_, appears to be extinct. Its leaves are much smaller than those of the living _L. asplenifolius_. The fossil species is known from a number of sites in Nevada and differs markedly from the abundant material in the Stewart Springs flora which represents a species (_L. cedrusensis_) with much larger leaves, closer to the living _asplenifolius_. _L. parvifolius_ may well have been a shrub rather than a small tree like _asplenifolius_, and is therefore listed under that category.

TABLE 1

Relations of Middlegate Species to Living Plants, Arranged According to the Usual Habit of the Taxa

FOSSIL SPECIES	COMPARABLE LIVING SPECIES: WESTERN NO. AMERICA	EASTERN NO. AMERICA	EASTERN ASIA
TREES (36)			
Conifers (10)			
Abies laticarpus	magnifica		
Abies scherri	bracteata		
Chamaecyparis cordillerae	extinct (aff. nootkatensis)		
Juniperus nevadensis	osteosperma		
Picea lahontense			polita
Picea sonomensis	breweriana		likiangensis
Pinus sturgisii	ponderosa		
Pseudotsuga sonomensis	menziesii		
Sequoiadendron chaneyi	giganteum		
Tsuga mertensioides	mertensiana		
Deciduous hardwoods (19)			
Acer middlegatensis		saccharinum	
Acer negundoides	negundo	negundo	henryi
Acer oregonianum	macrophyllum		
Acer scottiae			pictum
Acer tyrrelli	grandidentatum	saccharum	
Alnus largei			cremastogyne
Betula thor	papyrifera	papyrifera	japonica
Betula vera		lutea	
Diospyros oregonianum		virginiana	lotus
Fraxinus coulteri	oregona	americana	
Gymnocladus dayana		dioica	
Juglans nevadensis	californica		
Platanus dissecta	extinct (aff. racemosa)		
Platanus paucidentata	racemosa		
Populus bonhamii	balsamifera	balsamifera	
Populus cedrusensis	brandegeei		

- cont. -

TABLE 1, cont.

FOSSIL SPECIES	COMPARABLE LIVING SPECIES: WESTERN NO. AMERICA	EASTERN NO. AMERICA	EASTERN ASIA
Populus eotremuloides	hastata		adenopoda
Populus payettensis	angustifolia		
Populus pliotremuloides	tremuloides	tremuloides	
Evergreen sclerophylls (7)			
Arbutus prexalapensis	arizonica; others		
Cedrela trainii	mexicana		
Chrysolepis sonomensis	chrysophylla		
Lithocarpus nevadensis	densiflorus v. lanceolata		
Quercus hannibali	chrysolepis		
Quercus shrevoides	shrevei		
Quercus simulata	extinct (aff. chrysolepis)		
	SHRUBS/SMALL TREES (25)		
Deciduous (17)			
Acer nevadensis	diffusum		
Alnus harneyana	tenuifolia	incana	
Cercocarpus antiquus	betuloides aff.		
Cercocarpus eastgatensis	breviflorus		
Crataegus middlegatensis	chrysocarpa		
Crataegus pacifica			monogyna
Fraxinus millsiana	anomala		
Hydrangea ovatifolius			paniculata; heteromalla
Prunus moragensis	emarginata		
Robinia californica	neomexicana		
Salix owyheeana	hookeriana	bebbiana	
Salix pelviga	melanopsis		
Salix storeyana	lemmonii; argophylla		

- cont. -

TABLE 1, cont.

FOSSIL SPECIES	COMPARABLE LIVING SPECIES: WESTERN NO. AMERICA	EASTERN NO. AMERICA	EASTERN ASIA
Salix venosiuscula	caudata		
Salix wildcatensis	lasiolepis		
Sorbus idahoensis	scopulina	americana	acuparia; pohuashanensis
Styrax middlegatensis	californica		
Evergreen (8)			
Ceanothus precuneatus	cuneatus		
Chrysolepis convexa	sempervirens		
Heteromeles sonomensis	arbutifolia		
Lyonothamnus parvifolius	extinct (aff. asplenifolius)		
Mahonia macginitiei	aquifolium; piperiana		
Mahonia reticulata	insularis		
Mahonia simplex			lomariifolia
Mahonia trainii	nervosa		
AQUATIC HERBACEOUS PERENNIALS (3)			
Equisetum alexanderi	spp.	spp.	spp.
Nymphaeites nevadensis	Nymphaea spp.		
Typha lesquereuxi	latifolia	latifolia	latifolia

Attention is also directed to the leaves of Quercus shrevoides. The Nevada records appear to represent a distinct taxon, with leaves most similar to those produced by Q. shrevei in the sclerophyll forest belt of the outer Coast Ranges. Those trees produce leaves that are larger and not so hard and crisped as those formed by Q. wislizenii, which is a major component of the oak woodland-grass belt that surrounds the hot, dry interior valleys of California. The montane form, which lives under a wetter, more equable climate, is sufficiently distinct to deserve acceptance as a species. In any event, since the fossil leaves are like those produced in the areas where _Q. shrevei_ contributes to sclerophyll forest, _Q. shrevoides_ was probably a regular member of that forest which covered much of the Middlegate area.

These relations are paralleled by _Lithocarpus_, for its leaves are most similar to those formed by the species in the inner Coast Ranges and the central Sierra Nevada. In these areas, where it is a regular member of the sclerophyll forest and also reaches up into the lower margin of the mixed-conifer forest, the leaves are typically lanceolate and often entire, or with only a few small teeth; leaves formed by the species in the outer coastal region are regularly broader with numerous teeth, and are rarely entire. The interior form with slender leaves was recognized by Jepson (1910) as _forma lanceolata_. Ecologically, it prefers areas of more continental climate where sclerophyll vegetation dominates. By contrast, the broadleaved form is a regular member of the _Sequoia_ and _Pseudotsuga_ forests of coastal California and Oregon, where the climate is more equable.

As for the representation of taxa (Table 2), the leaves of _Quercus hannibali_ dominate the flora, accounting for nearly 85% of the collection. The next 3 most abundant species, _Quercus shrevoides_, _Cercocarpus antiquus_, and _Lithocarpus nevadensis_, are also sclerophylls and total another 7.55% of the sample. Among the 15 next most abundant taxa, all with more than 10 specimens, there are 4 species of _Acer_ and 1 each of _Salix_, _Betula_, _Populus_, and _Sequoiadendron_. All of these are typical of stream- or lake-border sites, and would thus be in a favorable position to contribute their structures to the accumulating record. The seeds of spruce (_Picea_) are also among this group, but its cones, needles,

TABLE 2

Quantitative Representation of Plant Structures in the Middlegate Flora*

SPECIES		NUMBER OF SPECIMENS	PERCENTAGE OF FLORA
Quercus hannibali		5,766	84.66%
leaves	5,763		
acorn cups	3		
Quercus shrevoides		252	3.67
Cercocarpus antiquus		141	2.05
Lithocarpus nevadensis		126	1.83
Typha lesquereuxi		94	1.37
Acer oregonianum		86	1.25
samaras	74		
leaves	12		
Populus bonhamii		40	.69
Salix storeyana		33	.48
Acer negundoides		33	.48
samaras	27		
leaflets	6		
Acer tyrrelli		28	.41
samaras	19		
leaves	9		
Picea sonomensis (samaras)		24	.35
Picea lahontense (samaras)		18	.26
Acer middlegatensis		16	.23
samaras	9		
leaves	7		
Arbutus prexalapensis		14	.20
Betula thor		14	.20
Lyonothamnus parvifolius		13	.19
Robinia californica		13	.19
leaflets	7		
pods	6		
Sequoiadendron chaneyi (branchlets)		12	.18
Abies laticarpus		11	.17
samaras	9		
needle	1		
cone scale	1		
Equisetum alexanderi (stems)		9	.13
Pinus sturgisii		9	.13
samaras	2		
needles	7		
Chrysolepis convexa		8	.12
Betula vera		7	.10
Cercocarpus eastgatensis		7	.10
Populus eotremuloides		7	.10
Juniperus nevadensis		7	.10
Quercus simulata		6	.09
Populus payettensis		6	.09

- cont. -

TABLE 2, cont.

SPECIES		NUMBER OF SPECIMENS	PERCENTAGE OF FLORA
Salix pelviga		6	.09
Fraxinus coulteri (samaras)		5	.07
Mahonia reticulata		5	.07
Fraxinus millsiana (samaras)		5	.07
Chamaecyparis cordillerae		4	.06
Nymphaeites nevadensis		4	.06
leaves	2		
rootstocks	2		
Crataegus middlegatensis		4	.06
Acer nevadensis		4	.06
samaras	3		
leaf	1		
Pseudotsuga sonomensis (samaras)		3	.04
Tsuga mertensioides (samaras)		3	.04
Chrysolepis sonomensis		3	.04
Acer scottiae		3	.04
samara	1		
leaves	2		
Cedrela trainii (samaras)		3	.04
Abies scherri		2	.03
samara	1		
needle	1		
Populus pliotremuloides		2	.03
Mahonia simplex		2	.03
Platanus paucidentata		2	.03
Heteromeles sonomensis		2	.03
Sorbus idahoensis		2	.03
Hydrangea ovatifolius		2	.03
Salix owyheeana		1	.01
Salix venosiuscula		1	.01
Salix wildcatensis		1	.01
Populus cedrusensis		1	.01
Alnus harneyana		1	.01
Alnus largei		1	.01
Juglans nevadensis		1	.01
Mahonia macginitiei		1	.01
Mahonia trainii		1	.01
Platanus dissecta		1	.01
Crataegus pacifica		1	.01
Gymnocladus dayana		1	.01
Prunus moragensis		1	.01
Ceanothus precuneatus		1	.01
Diospyros oregonianum		1	.01
Styrax middlegatensis		1	.01
Totals		6,882	100.93%

*Leaves, unless otherwise indicated.

or twigs were not encountered in the deposit. This implies that spruce was not near at hand, and that the light-winged seeds were probably carried by air or water currents to the site of accumulation. The fir *Abies laticarpus* is represented by 9 winged seeds, 1 needle, and 1 cone scale. Fir has a wider range of drought tolerance than spruce, and would descend to lower levels; this probably accounts for the somewhat better representation of its structures. *Arbutus* is also among the relatively more abundant taxa. It is commonly found in moister sites, and probably was a rather frequent member of the sclerophyll forest, together with others noted below.

The data in Table 2 indicate the relative development of plant communities near the fossil site. It seems obvious that if sclerophyll taxa dominate the record, a well-developed sclerophyll forest must have been near at hand. Likewise, if the remains of forest conifers are rare, they probably lived in a more distant area, reaching down valleys that carried cold-air drainage closer to the lake, where they provided local sites favorable for a few individuals. The belief expressed by Wolfe (1964, 1972), that a *Quercus*-*Chamaecyparis*-*Picea* forest dominated central Nevada in the Miocene, is patently incorrect. It finds no support in either the modern ecologic relations of the species or the fossil representation. The rare occurrence of *Chamaecyparis* in the numerous middle and late Miocene floras of Nevada, which are dominated by sclerophyll forest taxa, is consistent with its riparian distribution today. It descends from forested mountains into the lowlands, as on the Sacramento River near Castle Crags, where a thin stringer of scattered, rare trees that overhang the river would potentially contribute only occasional structures to an accumulating record. The adjacent plain and low hills are covered with oak-woodland and numerous sclerophyllous taxa. Such an occurrence parallels its record in the Miocene of Nevada. As presently known, *Chamaecyparis* was not an abundant species in the Miocene at most localities, for most floras include scarcely 10-12 small branchlets and these rarely exceed 2-3 cm in length. If Chamaecyparis were a codominant of the lowland forest in Nevada, numerous large branchlets with cones would be expected in the record. Furthermore, the other conifer forest taxa with which Chamaecyparis is regularly associated are also rare in these floras. This applies to Abies, Pinus, and Pseudotsuga, and especially to the Picea which Wolfe supposes was a codominant of the lowland forest. Picea is represented only by light winged seeds. Needles, twigs, and cones of Picea have not been recorded in the several floras known from the region, and the other conifers are also represented chiefly by winged seeds. Clearly, Picea and the other conifers lived on bordering, cooler slopes well away from the site of accumulation; otherwise their remains would be better represented. These floras are dominated by sclerophyll forest taxa; mixed-conifer forest was probably in the bordering hills, occupying cooler slopes from which some taxa descended to lower levels to contribute a few specimens to the deposits.

Vegetation

The nature of vegetation near the Middlegate site is suggested by an analysis of the flora in relation to modern forests. The Middlegate flora is composed of taxa whose nearest living allies contribute chiefly to the forests of California and bordering areas; only a few occur in the forests of the eastern United States, eastern Asia, or Mexico (Table 1). These diverse forests, which differ considerably in composition today, are segregates of the more generalized forests of the Miocene. In the western United States, those forests with taxa allied to the fossil flora are chiefly of two sorts, sclerophyll and mixed-conifer forest. In the eastern United States and eastern Asia, the living species allied to the fossil taxa contribute to mixed deciduous forests with numerous evergreens and deciduous hardwoods. In the Mexican sector, the allied living taxa are chiefly evergreen dicots that contribute to sclerophyll vegetation. Analysis of these forests allows us to infer the nature of the fossil plant communities and their distribution about the Middlegate basin.

Of the 61 woody taxa in the Middlegate flora, 51 have their nearest living relatives in the western United States and most of these occur in the sclerophyll and mixed-conifer forests of California. They account for over two-thirds of the fossil taxa and represent nearly 95% of the specimens studied. As noted earlier, one species, Quercus hannibalii, makes up 85% of our entire collection. This abundance probably was due chiefly to its restriction to moist valley sites, much as its nearest relative, Q. chrysolepis, occurs today. While this may account for the over-representation of its leaves in our sample, this oak was probably very prominent in the sclerophyll forest, and contributed also to the lower margin of the mixed-conifer forest, much as its equivalent does today.

Representation of the other sclerophylls in the area was undoubtedly higher than we might otherwise assume from the species frequency as judged from leaf counts. Regardless of their precise numerical representation, members of sclerophyll vegetation dominate the flora, and mixed-conifer forest taxa are less common to quite rare. Therefore, in order to evaluate conditions near the site, the modern sclerophyll forest is reviewed first, and then its nature in areas where it is adjacent to mixed-conifer forests. Communities in the region to the north or in more distant areas containing taxa similar to those in the Middlegate flora are then considered.

Sclerophyll forest, often referred to as oak-madrone or tanoak-madrone forest, has its optimum development under the moist, mild-winter climate of the Coast Ranges in central California, forming dense stands in the Santa Cruz and Santa Lucia mountains (Cooper, 1922). In this latter area, where it seems best developed, the forest has an elevation span of fully 760-915 m (2,500-3,000 ft.), from the woodland-grass ("savanna") of the *Quercus lobata-Q. douglasii-Pinus sabiniana* belt up to the patches of conifer forest on the summits of the range, where it is cooler and wetter. Farther north, its taxa are associated more regularly with members of the mixed-conifer (Sierran) forest, or with *Pseudotsuga*, *Sequoia*, and their associates on the coastward slopes. The fact that conifers and their usual associates are rarely represented in the Middlegate flora implies that the lowlands around the lake near the Middlegate site were covered primarily with sclerophyll forest, and that mixed-conifer forest was confined chiefly to higher levels. Species in the Coast Range sclerophyll forest that have allies in the fossil flora include *Abies bracteata*, °*Arbutus menziesii*, *Ceanothus cuneatus*, °*Cercocarpus betuloides*, *Chrysolepis chrysophylla*, *Heteromeles arbutifolia*, *Lithocarpus densiflorus*, *Mahonia aquifolium-piperiana*, *M. nervosa*, *Quercus chrysolepis*, and *Q. shrevei*. Streambank species in this zone with equivalent taxa in the flora include *Acer macrophyllum*, *A. negundo*, *Fraxinus oregona*, *Platanus racemosa*, °*Populus trichocarpa*, *Prunus emarginata*, *Salix hookeriana*, *S. lasiolepis*, and *S. melanopsis*.

Most of these species reach up to meet patches of the Sierran conifer forest in the Santa Lucia Mountains, and similar relations are apparent farther north in the Coast Ranges, though *Abies bracteata* and *Platanus racemosa* are not found there. Additional species associated with sclerophyll forest where it meets mixed-conifer forest include *Calocedrus decurrens*, *Pinus lambertiana*, *P. ponderosa*, and *Pseudotsuga menziesii*, as well as species of *Amelanchier*, *Ribes*, *Rosa*, and other common forest shrubs.

Views of the sclerophyll forest in the Santa Lucia Mountains shown on Plate 2, figures 1 and 2, include many of

°Ecologic equivalents of Middlegate taxa.

the important taxa noted above, especially *Abies bracteata*, *Arbutus menziesii*, *Lithocarpus densiflorus*, *Quercus chrysolepis*, and *Q. shrevei*, as well as *Pinus coulteri*, *P. lambertiana*, *P. ponderosa*, and their associates. The sclerophyll forest ranges farther south, reaching its southern limits on the upper slopes of the Santa Ynez Mountains and nearby areas along the Santa Barbara coastal strip. On Santa Cruz Island offshore are *Lyonothamnus asplenifolius*, which is allied to the fossil *L. parvifolius* that seems to be an extinct species, and *Mahonia insularis* (=*M. pinnata* var. *insularis*), which occurs there as a semi-vine up to 10 m long and seems to be the closest ally of the fossil *M. reticulata*; its nearest relatives are in central Mexico.

The occurrence of members of a *Sequoiadendron* forest in the flora gives us a basis for interpreting the environment in the forest that bordered the sclerophyll vegetation. Among species in the discontinuous groves in the Sierra Nevada are the following:

Common Woody Taxa

Abies concolor
*Abies shastensis (S)
Pinus lambertiana
Pinus ponderosa
*Pseudotsuga menziesii (N)
Calocedrus decurrens
*Sequoiadendron giganteum
Taxus breviflora (N)
*Populus tremuloides (C)
Salix laevigata
*Salix lasiolepis
*Salix lemmonii
*Salix melanopsis
Salix nuttallii
Alnus rhombifolia
*Alnus tenuifolia (C)
*Chrysolepis chrysophylla (C)
*Chrysolepis sempervirens
*Lithocarpus densiflora var. lanceolata (N)
Quercus breweri (S)
*Quercus chrysolepis
Quercus kelloggii
Corylus californica

°Mahonia sonnei (C)
Ribes nevadense
Ribes roezlii
*Amelanchier florida
°Cercocarpus betuloides (C,S)
Chamaebatia foliolosa
Holodiscus discolor
*Prunus emarginata
Rosa gymnocarpa
Rubus parviflorus
*Acer glabrum (C,N)
*Acer macrophyllum
Cornus californica
Cornus nuttallii
Ceanothus cordulatus
Ceanothus integerrimus
Rhamnus californica
Fremontodendron californicum (S)
Arctostaphylos patula
Arctostaphylos nevadensis
Rhododendron occidentale
Vaccinium parvifolium (N)
Symphoricarpos albus

- cont. -

Other Plants Below Bigtree Groves

Riparian woodland

*Acer negundo
*Platanus racemosa
Torreya californica
*Fraxinus oregona
Rhus diversiloba
Umbellularia californica
Vitis californica

Oak woodland/chaparral

Aesculus californica
*Ceanothus cuneatus
Arctostaphylos spp.
*Heteromeles arbutifolia
Cercis occidentalis
Quercus dumosa
°Quercus wislizenii
Styrax californica

Other Forest Species at Higher Levels

*Abies magnifica
Pinus contorta
*Tsuga mertensiana
°Pinus monticola
Quercus vaccinifolia
Ribes viscosissimum
Ribes montigenum
Salix spp.

*Represented by equivalent species in the fossil flora.
°Represented by allied species in the fossil flora.
N, C, S = chiefly northern, central, or southern.

Species in the Middlegate flora are similar to those found in both the northern and southern parts of Sequoiadendron distribution today. Plate 3, figure 2, shows a view of North Grove east of Foresthill at an elevation close to 1,520 m (5,000 ft.). Here Sequoiadendron is associated with Pseudotsuga menziesii and Lithocarpus, although the latter taxon there is the shrub variety L. echinoides. The tree form occurs nearby in the lower parts of the mixed-conifer forest. Other taxa in the North Grove area may be judged from the preceding list, which includes the commoner species in the Sierran groves. By contrast, the view seen in Plate 3, figure 1, shows the forest on the Tule River in the southern Sierra Nevada, where Sierra redwood is associated with

Quercus chrysolepis, Cercocarpus betuloides, and other sclerophylls with equivalents in the fossil flora, including Quercus wislizenii, allied to Q. shrevei, nearby. At slightly lower levels, a number of riparian taxa with correlatives in the flora are also present, notably Fraxinus oregona, Platanus racemosa, and several willows. Neither Lithocarpus nor Pseudotsuga occur in this area, being confined to areas farther north where precipitation is higher.

Other taxa in the flora have their nearest allies in northwestern California, notably Picea breweriana and Chamaecyparis nooktanensis in the Klamath-Siskiyou Mountains region. They occur there in the mixed-conifer forest associated with broadleaved sclerophyll taxa which have correlatives in the flora, notably Arbutus, Chrysolepis, and Quercus chrysolepis, as well as montane species of Abies, Picea, Pinus, Pseudotsuga, and their frequent associates. The species of Picea and Chamaecyparis are Tertiary relicts, having found a haven in this area where there is some summer rainfall and the more moderate summer temperatures alleviate the stress of the long dry season. Among the taxa in the Siskiyou Mountains associated with Picea and Chamaecyparis are the following in the area of Bear Basin Butte, situated 32 km (20 mi.) east of Crescent City at an elevation of 1,300 m (4,300 ft.):

Abies concolor
*Abies magnifica
*Picea breweriana
Pinus attenuata
Pinus lambertiana
*Pseudotsuga taxifolia
*Tsuga mertensiana
*Chamaecyparis nootkatensis
Calocedrus decurrens
Taxus breviflora
Salix nuttallii
Alnus sinuata
*Alnus tenuifolia
*Chrysolepis chrysophylla
*Chrysolepis sempervirens
Quercus breweri
*Quercus chrysolepis
Quercus vaccinifolia
Quercus sadleriana
*Mahonia aquifolium
*Mahonia nervosa

Amelanchier florida
°Crataegus douglasii
Prunus demissa
*Prunus emarginata
Holodiscus discolor
Rosa nutkana
Rubus parviflorus
Acer circinatum
°Acer glabrum
*Acer macrophyllum
Paxistima myrsinites
Ceanothus sanguineus
Ceanothus velutinus
Rhamnus purshiana
Cornus californica
Cornus nuttallii
°Arbutus menziesii
Arctostaphylos nevadensis
Arctostaphylos patula
Rhododendron californicum
Vaccinium parvifolium

Another area of import is the Sacramento River at Castle Crags State Park, just south of Dunsmuir. Here, in the ecotone between sclerophyll and mixed-conifer forest, are numerous taxa with comparable species in the flora, notably the following:

°Abies concolor
Pinus attenuata
Pinus lambertiana
Pinus ponderosa
*Pseudotsuga taxifolia
*Chamaecyparis lawsoniana
Calocedrus decurrens
Taxus brevifolia
*Populus tremuloides
*Populus trichocarpa
*Salix laevigata
*Salix lemmonii
Salix scouleriana
Alnus rhombifolia
*Alnus tenuifolia
°Betula occidentalis
*Chrysolepis chrysophylla
*Chrysolepis sempervirens
*Lithocarpus densiflora
*Quercus chrysolepis
Quercus garryana
Quercus kelloggii
Corylus californica
*Mahonia nervosa
Ribes roezlii
Ribes sanguineum
Amelanchier florida
Holodiscus discolor
Prunus demissa
*Prunus emarginata
Rosa gymnocarpa
Rubus parviflorus
Acer circinatum
°Acer glabrum
*Acer macrophyllum
Ceanothus integerrimus
Ceanothus velutinus
Rhamnus californica
Rhododendron occidentale

A few miles southwest, at an altitude near 1,525 m (5,000 ft.), most of these taxa range up to meet a subalpine forest composed of the following plants, as may be seen near Castle Lake and Castle Crags: **Abies magnifica*, **Picea breweriana*, *Pinus monticola*, *P. murrayana*, and **Tsuga mertensiana*.

Thirteen species in the Middlegate flora have their nearest descendants in areas with summer rainfall. These include *Populus bonhamii*, closely similar to *P. balsamifera*, which is scattered in the central and northern Rocky Mountains but occurs chiefly in the northeastern United States and Canada. In addition, *Betula thor* is allied to *B. papyrifera*, which ranges from eastern Washington-Idaho eastward into the northeastern United States. *Populus payettensis* is similar to *P. angustifolia*, which occurs chiefly in the central and southern Rocky Mountains, as do two additional taxa allied to those in the fossil flora: *Crataegus middlegatensis*, similar to *C. chrysocarpa*, and *Acer tyrrelli*, showing relationship to *A. grandidentatum* var. *brachypterum*, a western relative of the eastern sugar maples, *A. saccharum* and *A. leucoderme*. Two other taxa in the eastern United States with allies in the Middlegate flora are *Acer saccharinum* and *Diospyros virginiana*, allied to *A. middlegatensis* and *D. oregonianum*, respectively. There also are several species in the fossil flora with Asian affinities. These include *Alnus largei* (*cremastogyne*), *Crataegus pacifica* (cf. *monogyna*), *Hydrangea ovatifolius* (cf. *paniculata*), and *Picea lahontense* (cf. *polita*). Finally, there are two species that are largely of southern affinity: *Arbutus prexalapensis* resembles *A. arizonica* and other Mexican species of the genus, and *Cedrela trainii* shows relationship to *C. mexicana*. All of these taxa indicate the presence of some summer rainfall, as discussed further in the section on climate.

The preceding data indicate that the Middlegate area was dominated by a rich evergreen sclerophyll forest. A conifer-hardwood forest on bordering, moister slopes reached down as thin stringers toward the lake, contributing only a minority of structures to the accumulating record. It is evident that these Middlegate vegetation zones were more diverse in composition than those now living. This resulted in part from the presence of adequate summer rainfall that favored the intermingling of taxa whose nearest allies are now well separated geographically, or occur under different subclimates, as in California. In addition, these richer vegetation zones were also the result of more equable temperature. Since the thermal differences were reduced between the Tertiary vegetation zones, the taxa were more closely associated than are their descendants.

Table 3 lists the Middlegate flora according to the vegetation zones to which they most likely contributed. As shown there, riparian taxa extended into more than one vegetation zone, and also inhabited seepages and moist hollows on slopes as well. Some of the taxa listed under sclerophyll forest occurred also in the lower part of conifer-hardwood forest,

TABLE 3

Middlegate Taxa Grouped According to the Vegetation Types They Probably Represented

FOSSIL SPECIES	AQUATIC	LAKE-BORDER & RIPARIAN	SCLEROPHYLL FOREST & *SHRUBLAND	CONIFER-HARDWOOD FOREST
Equisetum alexanderi	x			
Nymphaeites nevadensis	x			
Typha lesquereuxi	x			
Acer negundoides		x	x	x
Alnus harneyana		x		x
Alnus largei		x		x
Betula thor		x		x
Fraxinus coulteri		x	x	x
Platanus dissecta		x		x
Platanus paucidentata		x	x	
Populus bonhamii		x		x
Populus cedrusensis		x	x	
Populus eotremuloides		x	x	x
Populus payettensis		x		x
Populus pliotremuloides		x		x
Salix owyheeana		x	x	x
Salix pelviga		x	x	x
Salix storeyana		x		x
Salix venosiuscula		x		x
Salix wildcatensis		x	x	x
Arbutus prexalapensis			X	x
Abies scherri			X	x
Chrysolepis convexa			X	x
Chrysolepis sonomensis			X	x
Lithocarpus nevadensis			X	x
Quercus hannibali			X	x
Quercus simulata			X	x
Quercus shrevoides			X	
Juglans nevadensis			X	
Ceanothus precuneatus			s	
Cercocarpus eastgatensis			s	
Cercocarpus antiquus			s	
Fraxinus millsiana			s	
Lyonothamnus parvifolius			s	
Mahonia reticulata			s	
Mahonia trainii				x
Robinia californica			s	
Styrax middlegatensis			s	
Abies laticarpus				X
Chamaecyparis cordillerae		x		X
Picea lahontense				X
Picea sonomensis				X
Pinus sturgisii				X

- cont. -

TABLE 3, cont.

FOSSIL SPECIES	AQUATIC	LAKE-BORDER & RIPARIAN	SCLEROPHYLL FOREST & *SHRUBLAND	CONIFER-HARDWOOD FOREST
Pseudotsuga sonomensis				X
Tsuga mertensioides				X
Juniperus nevadensis			s	
Sequoiadendron chaneyi		x		X
Acer oregonianum		x	x	x
Acer tyrrelli		x	x	x
Acer nevadensis		x		x
Acer middlegatensis				x
Acer scottiae				x
Betula vera				x
Cedrela trainii				x
Diospyros oregonianum				x
Gymnocladus dayana				x
Crataegus pacifica				x
Crataegus middlegatensis		x		x
Mahonia macginitiei			x	x
Mahonia simplex				x
Heteromeles sonomensis			x	x
Prunus moragensis		x		x
Sorbus idahoensis		x		x
Hydrangea ovatifolius				x

*X = dominant

x = also in understory of lower conifer-hardwood forest

s = forms local shrub patches

where they are denoted by a small "x." This indicates their subordinate position in the forest canopy as compared with their dominance in the sclerophyll forest, marked by a large "X." The shrub taxa in the sclerophyll vegetation are denoted by a small "s" to indicate that they may also have contributed to local shrub communities ("chaparral") wherever thin soil and dry southern exposures favored their presence, chiefly in seral relation.

Table 3 shows that more numerous taxa are listed for conifer-hardwood forest than for sclerophyll forest. As noted earlier, the low representation of remains of conifer-hardwood forest taxa implies that they lived chiefly upslope from the lake, away from the sclerophyll forest that dominated the warm, south-facing slopes along the lake shore. In such a setting, the conifer-hardwood forest taxa probably extended down valleys with cold-air drainage and thus contributed fewer structures to the record. Obviously, if conifer-hardwood forest taxa had lived closer to the shore area, a more complete record of their remains would have been recovered, including twigs, cones, and needles of the conifers, as well as more numerous leaves, fruits, and seeds of the associated dicots. Inasmuch as 11 of them are quite rare, except for some of the more widely-ranging riparian taxa of *Populus* and *Salix*, it seems reasonable to conclude that the conifer-hardwood forest taxa were farther removed from the area of plant accumulation than the dominant sclerophylls.

THE EASTGATE FLORA

Composition

As now known, the Eastgate flora includes 55 species distributed among 5,805 specimens. The flora is represented by 14 conifers, distributed among 9 genera, and 41 angiosperms, represented by 26 genera. There are 13 new species, distributed in *Acer*, *Cercocarpus* (2), *Chamaecyparis*, *Eugenia*, *Larix*, *Lithocarpus*, *Mahonia*, *Populus*, *Quercus*, *Salix*, *Sorbus*, and *Sparganium*.

Systematic List of Species

Pinaceae

- Abies concoloroides Brown
- Abies laticarpus MacGinitie
- Larix cassiana Axelrod
- Larix nevadensis Axelrod, new species
- Picea lahontense MacGinitie
- Picea sonomensis Axelrod
- Pinus alvordensis Axelrod
- Pinus balfouroides Axelrod
- Pinus sturgisii Cockerell
- Pseudotsuga sonomensis Dorf
- Tsuga mertensioides Axelrod

- cont. -

Cupressaceae
- Chamaecyparis cordillerae Edwards and Schorn, new species
- Juniperus nevadensis Axelrod

Taxodiaceae
- Sequoiadendron chaneyi Axelrod

Typhaceae
- Typha lesquereuxi Cockerell

Sparganiaceae
- Sparganium nevadense Axelrod, new species

Salicaceae
- Populus bonhamii Axelrod, new species
- Populus cedrusensis Wolfe
- Populus eotremuloides Knowlton
- Populus payettensis (Knowlton) Axelrod
- Populus pliotremuloides Axelrod
- Salix desatoyana Axelrod, new species
- Salix pelviga Wolfe
- Salix storeyana Axelrod
- Salix venosiuscula Smith

Betulaceae
- Betula thor Knowlton
- Betula vera Brown

Fagaceae
- Chrysolepis convexa (Lesq.) Axelrod, new comb.
- Lithocarpus nevadensis Axelrod, new species
- Quercus hannibali Dorf
- Quercus shrevoides Axelrod, new species
- Quercus simulata Knowlton

Berberidaceae
- Mahonia macginitiei Axelrod, new species
- Mahonia reticulata (MacGinitie) Brown
- Mahonia simplex (Newberry) Arnold

Ceratophyllaceae
- Ceratophyllum praedemersum Ashlee

Nymphaeaceae
- Nymphaeites nevadensis (Knowlton) Brown

Grossulariaceae
- Ribes stanfordianum Dorf

Rosaceae
- Amelanchier grayi Chaney
- Cercocarpus eastgatensis Axelrod, new species
- Cercocarpus ovatifolius Axelrod, new species
- Crataegus newberryi Cockerell
- Heteromeles sonomensis (Axelrod) Axelrod, new comb.
- Lyonothamnus parvifolius (Axelrod) Wolfe

- cont. -

Prunus chaneyi Condit
Sorbus idahoensis Axelrod, new species

Leguminosae
Robinia californica Axelrod

Aceraceae
Acer nevadensis Axelrod, new species
Acer oregonianum Knowlton
Acer tyrrelli Smiley

Hippocastanaceae
Aesculus preglabra Condit

Rhamnaceae
Rhamnus precalifornica Axelrod

Myrtaceae
Eugenia nevadensis Axelrod, new species

Ericaceae
Arbutus prexalapensis Axelrod

Oleaceae
Fraxinus coulteri Dorf

Table 4 presents a numerical count of Eastgate specimens. The most abundant remains are the leaves of 2 sclerophylls, Quercus hannibali and Lithocarpus nevadensis, which together account for 4,058 specimens, or nearly 70% of the sample of 5,805 specimens. Of notable interest is the high representation of Sierra redwood, for its remains, which include several heavy cones and large branchlets, are sufficiently common to make it third in abundance. Of further significance is the abundance of conifers, with 14 species distributed among 827 specimens making up 14.3% of the total sample. The Eastgate flora has a greater representation of conifers than any other Miocene flora previously described from this region. Only the nearby Buffalo Canyon flora of closely similar age has more numerous conifer remains.

These numerical data obviously cannot provide an index to the abundance of the taxa in the vegetation of the area, as some have mistakenly asserted. To judge from the habitats occupied by their living descendants, the most common species

TABLE 4

Quantitative Representation of Plant Structures in the Eastgate Flora*

SPECIES		NUMBER OF SPECIMENS	PERCENTAGE OF FLORA
Quercus hannibali		2,568	44.10%
leaves	2,563		
acorn cups	5		
Lithocarpus nevadensis		1,497	25.83
leaves	1,495		
acorn cups	2		
Sequoiadendron chaneyi		277	4.78
leafy twigs	273		
cones	4		
Betula thor		250	4.31
leaves	235		
seeds	5		
aments	10		
Picea lahontense (samaras)		224	3.87
Typha lesquereuxi		160	2.76
Cercocarpus ovatifolius		104	1.79
Acer tyrrelli		90	1.55
samaras	4		
leaves	86		
Picea sonomensis (samaras)		85	1.47
Pseudotsuga sonomensis (samaras)		57	1.00
Pinus sturgisii		52	.90
samaras	45		
needles	7		
Mahonia reticulata		52	.90
Salix pelviga		48	.83
Tsuga mertensioides (samaras)		38	.66
Chamaecyparis cordillerae		29	.50
leafy twigs	27		
cones	2		
Cercocarpus eastgatensis		28	.48
Populus cedrusensis		24	.41
Pinus alvordensis (samaras)		22	.38
Sorbus idahoensis		22	.38
Populus eotremuloides		15	.26
Abies laticarpus		14	.24
samaras	13		
needle	1		
Mahonia simplex		12	.21
Abies concoloroides		11	.19
samaras	7		
needles	3		
cone scale	1		

- cont. -

TABLE 4, cont.

SPECIES	NUMBER OF SPECIMENS	PERCENTAGE OF FLORA
Populus pliotremuloides	10	.17
Lyonothamnus parvifolius	9	.16
Quercus simulata	8	.14
Amelanchier grayi	8	.14
Larix nevadensis	7	.12
Aesculus preglabra	7	.12
Arbutus prexalapensis	7	.12
Quercus shrevoides	7	.12
Populus payettensis	6	.10
Salix storeyana	5	.09
Pinus balfouroides (winged seeds)	4	.07
Juniperus nevadensis	4	.07
Populus bonhamii	4	.07
Heteromeles sonomensis	4	.07
Chrysolepis convexa	4	.07
Larix cassiana	3	.05
Nymphaeites nevadensis	3	.05
leaf 1		
rootstock 2		
Ribes stanfordianum	3	.05
Prunus chaneyi	3	.05
Acer nevadensis	2	.03
Acer oregonianum (samaras)	2	.03
Betula vera	2	.03
Ceratophyllum praedemersum	2	.03
Robinia californica	2	.03
Salix desatoyana	2	.03
Sparganium nevadense	2	.03
Salix venosiuscula	1	.02
Mahonia macginitiei	1	.02
Crataegus newberryi	1	.02
Eugenia nevadensis	1	.02
Fraxinus coulteri (samara)	1	.02
Rhamnus precalifornica	1	.02
Totals	5,805	99.96%

*Leaves, unless otherwise indicated

(*Quercus hannibali*, *Lithocarpus nevadensis*) not only covered hillslopes, but also occupied the borders of the lake and streams, where they formed dense sclerophyllous woodlands associated with birch, willow, cottonwood, and other taxa of generally hydric requirements. As a result, they were in a position to contribute more numerous structures to the accumulating record than were trees and shrubs inhabiting the adjacent slopes. Thus their high representation in the flora only reflects their occurrence in sites more favorable for preservation. The general nature of vegetation in the adjacent area can be inferred more reliably if we reduce the representation of the dominants that inhabited these hydric-border sites. This also seems appropriate because the area they occupied was so small, in terms of the adjacent slopes and plains, that they do not represent regional vegetation. If we reduce the specimens representing the 2 dominant taxa by only half, then the 14 conifers which had accounted for 14.3% of the total collection rise to 21.5%. This probably is more nearly representative of forest composition in areas only a few tens of meters removed from the lake margin. If we also consider the other species in the flora that lived with them, it is evident that the Eastgate flora represents a rich conifer-hardwood forest to which sclerophylls (*Arbutus*, *Lithocarpus*, *Quercus*) also contributed. In addition, they dominated restricted warmer slopes at lower levels, where they intermingled with *Acer*, *Betula*, *Populus*, *Salix*, and other near-hydric taxa.

The probable life-form or habit represented by the Eastgate plants may be inferred from that of living species most similar to them (Table 5). From this standpoint, the flora includes 31 trees, distributed among 14 conifers, 11 deciduous hardwoods, and 6 evergreen sclerophylls. There are 20 shrubs or small trees, of which 13 are deciduous and 7 evergreen. The remaining 4 species are aquatics: cattail (*Typha*), waterlily (*Nymphaeites*), water milfoil (*Ceratophyllum*) and bur-reed (*Sparganium*).

TABLE 5

Relations of Eastgate Species to Living Plants, Arranged According to the Usual Habit of the Taxa

FOSSIL SPECIES	COMPARABLE LIVING SPECIES WESTERN NO. AMERICA	EASTERN NO. AMERICA	EASTERN ASIA
	TREES (31)		
Conifers (14)			
Abies concoloroides	concolor		
Abies laticarpus	magnifica		
Chamaecyparis cordillerae	lawsoniana		
Juniperus nevadensis	osteosperma		
Larix cassiana			potaninii
Larix nevadensis	occidentalis		
Picea lahontense			polita
Picea sonomensis	breweriana		likiangensis
Pinus alvordensis	murrayana		
Pinus balfouroides	balfouriana		
Pinus sturgisii	ponderosa		
Pseudotsuga sonomensis	menziesii		
Sequoiadendron chaneyi	giganteum		
Tsuga mertensioides	mertensiana		
Deciduous hardwoods (11)			
Acer oregonianum	macrophyllum		
Acer tyrrelli	grandidentatum		
Aesculus preglabra		glabra; pavia	
Betula thor	papyrifera	papyrifera	japonica
Betula vera		lutea	
Fraxinus coulteri	oregona	americana	
Populus bonhamii	balsamifera	balsamifera	
Populus cedrusensis	brandegeei		
Populus eotremuloides	hastata		adenopoda
Populus payettensis	angustifolia		
Populus pliotremuloides	tremuloides	tremuloides	

- cont. -

TABLE 5, cont.

FOSSIL SPECIES	COMPARABLE LIVING SPECIES WESTERN NO. AMERICA	EASTERN NO. AMERICA	EASTERN ASIA
<u>Evergreen sclerophylls</u> (6)			
Arbutus prexalapensis	arizonica; others		
Eugenia nevadensis	(Mexico, spp.)		
Lithocarpus nevadensis	densiflorus var. lanceolata		
Quercus hannibali	chrysolepis		
Quercus shrevoides	shrevei		
Quercus simulata	extinct (aff. chrysolepis)		
	SHRUBS (20)		
<u>Deciduous</u> (13)			
Acer nevadensis	diffusum		
Amelanchier grayi	pallida; florida		
Cercocarpus eastgatensis	breviflorus; paucidentatus		
Cercocarpus ovatifolius	blancheae		
Crataegus newberryi			pinnatifida
Prunus chaneyi	demissa	virginiana	japonica
Ribes stanfordianum	nevadense		
Robinia californica	neomexicana		
Salix desatoyana		nigra	
Salix pelviga	melanopsis		
Salix storeyana	lemmonii		
Salix venosiuscula	caudata		
Sorbus idahoensis	°scopulina	°americana	aucuparia; pohuashanensis
<u>Evergreen</u> (7)			
Chrysolepis convexa	sempervirens		
Heteromeles sonomensis	arbutifolia		
Lyonothamnus parvifolius	asplenifolius aff.		
Mahonia macginitiei	aquifolium; pinnata; piperiana		

- cont. -

TABLE 5, cont.

FOSSIL SPECIES	COMPARABLE LIVING SPECIES WESTERN NO. AMERICA	EASTERN NO. AMERICA	EASTERN ASIA
Mahonia reticulata	insularis		
Mahonia simplex			lomariifolia
Rhamnus precalifornica	californica		
AQUATIC HERBACEOUS PERENNIALS (4)			
Ceratophyllum prademersum	demersum	demersum	demersum
Nymphaeites nevadensis	Nymphaea spp.	spp.	spp.
Sparganium nevadense	spp.	spp.	spp.
Typha lesquereuxii	latifolia	latifolia	latifolia

°Generally allied to species.

Of the 51 woody plants in the Eastgate flora, 42 have close analogues in the western United States. Most are in California, where they contribute to the Sierran mixed-conifer forest and to sclerophyll forest. Three of them not now in California, _Betula papyrifera_, _Larix occidentalis_, and _Populus balsamifera_, live east of the Cascades in northern Oregon, Washington, Idaho, and adjacent Canada, where they are in a mixed-conifer forest composed of _Pinus monticola_, _P. ponderosa_, _Pseudotsuga menziesii_, and their usual associates. _Acer diffusum_, _A. grandidentatum_, and _Populus angustifolia_ are common along streambanks in the central to southern Rocky Mountains, where they occur with _Abies concolor_, _Pinus ponderosa_, and _Pseudotsuga menziesii_, extending down into the adjacent piñon-oak woodland zone. _Arbutus arizonica_ occurs in the southwestern United States and northern Mexico, inhabiting the upper part of the rich evergreen sclerophyll forest and extending up to the lower margin of _Pinus ponderosa_ forest.

Of the 9 woody species in the eastern United States that show relationship to the fossils, 6 also have allies in the western United States. They are members of widely distributed phylads in the genera *Betula*, *Fraxinus*, *Populus*, *Prunus*, and *Sorbus*. The other 3 species, *Aesculus preglabra*, *Betula vera*, and *Salix desatoyana*, have their nearest descendants restricted to the eastern United States. Of the 9 species listed for eastern Asia that show relationship to the Eastgate taxa, 6 are in the widely distributed genera *Betula*, *Mahonia*, *Picea*, *Prunus*, *Populus*, and *Sorbus* that occur also in the western or eastern United States. Only a larch (*Larix*) spruce (*Picea polita*), the hawthorn (*Crataegus pinnatifida*), and mahonia (*Mahonia lomariifolia*) do not have closely related species in North America today. The only genus in the flora that is not represented in one of these three regions is *Eugenia*, which ranges from central Mexico southward. It is common in the Eocene Green River and Florissant floras of Colorado, but rare at Eastgate. *Eugenia* also occurs in 3 other middle to lower Miocene floras (Buffalo Canyon, Golddyke Road, and Carson Pass) of western Nevada and adjacent California, and it is also rare in them, being represented by only 1 or 2 specimens. Clearly, it was a relict in these younger floras.

The rarity of genera at Eastgate that now live only in the eastern United States or Asia contrasts with the record provided by floras in regions farther north, as in the 49-Camp flora of northwestern Nevada (LaMotte, 1936; Chaney, 1959, p. 117), the Mascall of central Oregon (Chaney, 1959, p. 16-17), the Succor Creek of the Oregon-Idaho border (Chaney, 1959, p. 114), or the Spokane and Grand Coulee "Latah" floras of eastern Washington (see lists in Chaney, 1959, p. 109-112), and the Clarkia of northern Idaho (Smiley and Rember, 1981). In those floras, genera no longer native to the western United States are not only well represented, some of them formed subdominants of those forests. Among the genera recorded are *Ailanthus*, *Carpinus*, *Castanea*, *Cedrela*,

Diospyros, Exbucklandia, Fagus, Halesia, Hamamelis, Ilex, Liriodendron, Machilis, Nyssa, Ostrya, Persea, Pterocarya, Sassafras, Sophora, Taxodium, Tilia, Ulmus, and Zelkova--none of which occur at Eastgate. In addition, the northern floras have more numerous species of Acer, Alnus, Betula, Crataegus, Fraxinus, Mahonia, Populus, Prunus, Quercus, and others whose nearest relatives are either in the eastern United States or eastern Asia where there is ample summer rainfall, but not in the western United States where summer precipitation is largely absent (California) or deficient (Oregon, Washington), apart from coastal summer fog which adds moisture to the soil. The rarity of these taxa in the Eastgate flora is also apparent when floras of similar age in the nearby region are examined. They are more common in the Buffalo Canyon flora, situated 14 km (9 mi.) southeast of the Eastgate locality, where species of Acer, Betula, Carya, Eugenia, Juglans, Mahonia, Populus, Robinia, Ulmus, and Zelkova occur, some of them (Betula, Carya, Ulmus, Zelkova) represented by numerous specimens. Furthermore, the younger (16 m.y.) Fingerrock flora from a locality 128 km (80 mi.) south has species of Carya, Diospyros, Populus, Gymnocladus, Ulmus, and Zelkova and several woody legumes--all now found only in regions with ample summer rainfall. As discussed below, these differences in composition may reflect diverse local topographic-climatic conditions, though other factors not yet recognized may also be responsible.

Our large collection has not yielded any specimens of the broadleaved deciduous oaks that are so common in many of the Miocene floras of the Pacific Northwest, notably Quercus columbiana, eoprinus, merriami, prelobata, or pseudolyrata. As noted earlier (Axelrod, 1964, p. 25), broadleaved deciduous oaks occur in greatest diversity and abundance in Miocene floras that lived at relatively low altitudes, under mild temperate to warm temperate climate (e.g., Latah, Clarkia, Eagle Creek, Mascall, Mollala, Temblor floras). They are less abundant, or absent, in floras from higher levels where cool temperate climate prevailed (e.g., Blue Mountains, Trapper Creek, Trout Creek floras). This marked decrease in diversity is correlated with a shift from deciduous hardwood forest to conifer-hardwood forest, as may be seen today in the Great Lakes area, in New England, or along a transect up into the Great Smoky Mountains. Since the Eastgate flora is a rich conifer-hardwood forest, it is understandable that broadleaved deciduous oaks would be rare or absent. This is consistent with their representation in Miocene floras of the nearby region. Broadleaved oaks are not present in the nearby Buffalo Canyon flora, which has numerous conifers. However, one species (*Q. pseudolyrata*) is well represented in the Lower Fingerrock flora that lived at a lower, warmer level 128 km (80 mi.) south, where conifers are not so abundant. To the west, broadleaved deciduous oaks are well represented in the Temblor flora from sea level in central California, where climate was much warmer than at Eastgate (Renney, 1972). Clearly, if broadleaved deciduous oaks inhabited the Middlegate basin they must have been very rare, not only at Eastgate but at Middlegate as well.

To summarize, the Eastgate flora of more than 5,800 specimens has 55 species distributed among 31 trees (14 conifers, 17 dicots), 20 shrubs or small trees, and 4 herbaceous aquatic perennials. Many of the fossil plants have their nearest living analogues in the conifer forests and bordering broadleaved sclerophyll forest of California. A few additional taxa occur in the Pacific Northwest and in the southwestern United States or Mexico. In contrast to floras of similar and younger age in regions to the north, taxa with relatives in the forests of the eastern United States or eastern Asia are notably rare and poorly represented.

Vegetation

Most Eastgate fossils resemble living species, many of which are still associated today and contribute to unique forest communities. The vegetation of the Eastgate area may be reconstructed, therefore, by reference to regions where groups of analogous species contribute to modern vegetation. Many Eastgate species have close relatives in the Sierra Nevada, where they make up the Sierra redwood forest, or are members of vegetation zones adjacent to it, either fir-subalpine or sclerophyll forest. Sierra redwood forest is distributed discontinuously through the central and southern parts of the range, finding optimum development at elevations ranging from 1,370-2,280 m (4,500-7,500 ft.). In Sequoia National Park, the forest groves include the following species that have close analogues in the Eastgate flora:

Abies concolor	Quercus chrysolepis
°Abies shastensis	Mahonia pinnata
Pinus murrayana	°Platanus racemosa
Pinus ponderosa	Amelanchier florida
Sequoiadendron giganteum	Ribes nevadense
Populus tremuloides	°Cercocarpus betuloides
Populus hastata	Prunus demissa
Salix lemmonii	°Sorbus scopulina
Salix melanopsis	°Acer glabrum
Chrysolepis sempervirens	Acer macrophyllum
Lithocarpus densiflorus	Rhamnus californica
	°Arbutus menziesii

°Allied to the fossil species, but not the closest modern equivalents.

Some of these grow close to fir-subalpine forest near the north margin of Giant Forest, at Lodgepole, at an elevation of 2,040 m (6,700 ft.). Here, subalpine forest is on the north-facing slope of the deep canyon of Marble Fork of the Kaweah River. Cold-air drainage from higher elevations enables *Pinus murrayana* and *P. monticola* to grow here close to Sierra redwood, and *Tsuga mertensiana* has its southern occurrence in the range at somewhat higher elevations in this drainage basin. Other plants in the area with analogues in the flora include *Acer glabrum* and *Sorbus scopulina*. The juxtaposition of fir-subalpine forest species with Sierran mixed-conifer forest is not unique, but occurs regularly in areas of favorable terrain. For instance, in the American

River canyon above Strawberry, Abies magnifica, Pinus monticola, P. murrayana, Acer glabrum, Sorbus scopulina, and Populus tremuloides reach lower altitudes on the north-facing slope. In this area (elev. 1825 m; 6,000 ft.) they are near Pinus ponderosa, P. lambertiana, Abies concolor, Calocedrus decurrens, Quercus kelloggii, Chrysolepis sempervirens, and their usual associates.

These occurrences might suggest that the subalpine forest species at Eastgate had a similar distribution, reaching down to lower levels on cooler north-facing slopes in canyons with cold-air drainage. Such a distribution would also be favored by the setting of the north-trending Eastgate Hills. Since the canyons faced west, they would tend to funnel prevailing summer breezes off the cooler lake surface and up the canyons, lowering temperature there, thus favoring the occurrence of upland conifers at lower elevations and nearer the lake. However, the upper margin of sclerophyll forest and the lower margin of subalpine forest are separated by fully 900 m (3,000 ft.) in the Sierra Nevada today. There certainly is no geologic evidence to support the notion that the Eastgate Hills stood 900 m (3,000 ft.) above the lakeshore, or that numerous fossil plants from such a summit region would survive transport into the fossil deposit (cf. Burrows, 1980). Clearly, some other factor must have enabled taxa of the subalpine forest belt to enter the record in rather large numbers.

As discussed elsewhere (Axelrod, 1976a), trees that are confined chiefly to the fir-subalpine forest in the Sierra Nevada also enter mixed-conifer forest in the Klamath-Siskiyou region and the southern Cascades. This is the result of more favorable conditions there, notably a longer rainy season (October into June, as compared with November into April) and hence more soil moisture during the critical dry season. Also, there is more cloudiness and rainfall in summer, which lowers the evaporation rate, and hence the stress of wilting on seedlings is reduced. Further, since extreme summer temperatures are not so high, they favor a lower evaporation rate and reduced water-stress during the crucial period of seedling establishment. Inasmuch as Abies lasiocarpa, A. magnifica or shastensis, Picea engelmannii, P. breweriana, Pinus monticola, Tsuga mertensiana, and others frequently extend down into the mixed-conifer forest in the Klamath-Siskiyou region, and also in the southern Cascades, we may infer that under a Miocene climate of wet summers and mild temperature fossil relatives of these taxa probably were regular members of a mixed-conifer forest richer than any now living. This is precisely what the record shows. The fossil conifers are sufficiently abundant that they could scarcely have been transported from subalpine forests in mountains 900-1,200 m (3,000-4,000 ft.) above the fossil basins, from sites fully 16 km (10 mi.) or more distant, as others have inferred (Chaney, 1959; Graham, 1965; Becker, 1961). On the contrary, the evidence suggests that they probably were regular members of both conifer-hardwood and fir-subalpine forests, and were restricted gradually to the latter zone in the Sierra Nevada as a montane mediterranean climate of progressively drier summers spread over the region during the Quaternary (Axelrod, 1976a, fig. 4).

In the northern part of its area, Sierra redwood lives with species that do not occur with it farther south. This probably is because precipitation decreases southward and is also less effective there, owing to warmer summers and higher evaporation. At North Grove, at 1,520 m (5,000 ft.) in the area 32 km (20 mi.) east of Foresthill, Sierra redwood is associated with Pseudotsuga menziesii and Lithocarpus densiflorus. Neither of these occur with Sequoiadendron elsewhere in the range, though Pseudotsuga is close to it just outside the North Calaveras Grove, and also occurs slightly below the Mariposa Grove. Woody plants at North Grove (Plate 3, fig. 2), or in the valley of Frasier Creek, a mile north, that have analogues in the Eastgate flora include:

Abies concolor
Pinus ponderosa
Pseudotsuga menziesii
Sequoiadendron giganteum
Salix lemmonii
Chrysolepis sempervirens
°Lithocarpus densiflorus var. echinoides
°Quercus vaccinifolia
Amelanchier pallida (=alnifolia)
°Sorbus scopulina
Ribes nevadense
Rhamnus californica

°Shrub forms of *L. densiflorus* and *Q. chrysolepis*.

Woody plants in this area not now known to have fossil representatives in the Eastgate flora include:

Pinus lambertiana	Cornus californica
Calocedrus decurrens	Cornus nuttallii
Salix nuttalli	Arctostaphylos patula
Alnus rhombifolia	Arctostaphylos nevadensis
Alnus tenuifolia	Leucothoe davisiae
Quercus kelloggii	Rhododendron occidentale
Ribes roezlii	Vaccinium parvifolium
Ribes monteginum	Garrya fremontii
Spiraea densiflora	Ceanothus integerrimus
Prunus emarginata	Ceanothus cordulatus
Rosa spp.	Symphoricarpos albus

Two of these (*Alnus tenuifolia*, *Prunus emarginata*) have correlative species (*A. harneyana*, *P. moragensis*) in the Middlegate flora, situated on the northwest shore of the basin 8 km (5 mi.) distant. It seems likely that some of the others may have had allied species in the Middlegate basin.

Quercus chrysolepis, whose fossil analogue *Q. hannibali* dominates the flora, forms dense woodlands on warmer slopes at moderate levels below North Grove. Madrone (*Arbutus menziesii*) is also present there, though the fossil appears to more nearly resemble living species in the southwestern United States and Mexico. *Acer macrophyllum*, common along stream borders and in moist swales in the Sierran mixed-conifer forest, occurs at a slightly lower elevation in the region near North Grove. It is also common with *Sequoiadendron* in the South Calaveras Grove, 50 miles southeast. *Fraxinus oregona*, which shows relationship to the fossil *F.*

coulteri, also occupies riparian sites at lower levels in the region.

There are other species in the Eastgate flora whose nearest relatives occur elsewhere in the mixed-conifer forest, but not in the Sierra Nevada. Among these, *Picea breweriana* and *Chamaecyparis lawsoniana* live in the Siskiyou and Klamath mountains of northwestern California, ranging up into the ecotone between mixed-conifer and subalpine forest. The following species with equivalents in the Eastgate flora are associated with them in the Siskiyou Mountains:

Abies concolor	Mahonia aquifolium
Abies magnifica	°Mahonia nervosa
Pinus balfouriana	Mahonia pinnata
Pinus murrayana	Ribes nevadense
Pinus ponderosa	Amelanchier pallida (=alnifolia)
Tsuga mertensiana	Prunus demissa
Populus tremuloides	°Sorbus sitchensis
Salix lemmonii	°Acer glabrum
Salix melanopsis	Acer macrophyllum
Chrysolepis sempervirens	Rhamnus californica
Lithocarpus densiflorus	°Arbutus menziesii

°Allied species, or ecologic equivalents.

The absence of *Chamaecyparis* and *Picea* in the Sierra Nevada today is probably due to the lower summer rainfall and warmer summers that result in less effective moisture. By contrast, summer rains total 75-100 mm (3-4 in.) monthly in the mountains over the Oregon-California border area where *Chamaecyparis* and *Picea* now live, and increase considerably in adjacent Oregon (Froelich, McNabb, and Gaweda, 1982). These conifers may have inhabited the Sierra Nevada well into the Pleistocene, being eliminated there during the Xerothermic period. As outlined elsewhere (Axelrod, 1966b, p. 42-44), the warmer, drier climate 8,000 to 6,000 years ago greatly restricted forests in the region, and accounts for many relict and disjunct distributions. That *Chamaecyparis* and *Picea* may have been in the northern half of the Sierra up to that time is indicated by the composition of Pleistocene pollen floras from the Chagoopa surface on the Kern Plateau in the southern Sierra Nevada (Axelrod and Ting, 1961). Those floras have taxa that now occur chiefly in coastal northern California and southern Oregon (e.g., *Abies grandis*, *Chamaecyparis lawsoniana*, *Salix hookeriana*, *Tsuga heterophylla*), or also in the northern half of the Sierra (e.g., *Acer circinatum*, *Chrysolepis chrysophylla*, *Pseudotsuga menziesii*, *Taxus brevifolia*). A study of 4 post-Wisconsin pollen floras from the middle and higher parts of the central Sierra Nevada did not reveal the presence of either *Picea* or *Chamaecyparis* (Adam, 1967); but if their distribution was relict and local at this time, as postulated here, they might not be recorded at or near these sites. Critical in this regard is a more recent paper by Adam (1973) that reports *Picea* grains near Tahoe City, in rocks older than 1.9 m.y., in the late Pliocene.

Three Eastgate species have modern equivalents in the conifer forest of eastern Washington and adjacent Idaho, *Betula papyrifera*, *Larix occidentalis*, and *Populus balsamifera*. Summer rainfall in that area totals 150 mm (6 in.) or more, an amount that seems to coincide with their southern distribution. The following plants with close equivalents in the Eastgate area occur with larch and birch in Idaho-Washington-Oregon:

Abies concolor	Salix caudata
°Abies procera (=nobilis)	°Salix melanopsis
°Pinus monticola	Mahonia aquifolium
Pinus murrayana	°Mahonia nervosa
Pinus ponderosa	Ribes nevadense
Pseudotsuga menziesii	Amelanchier florida
Tsuga mertensiana	Prunus demissa
Populus angustifolia	°Sorbus scopulina
Populus tremuloides	°Acer glabrum
Populus hastata	

°Allied taxa, or ecologic equivalents.

Only a few species in the Eastgate flora are not represented by equivalent taxa in the Pacific states. Three have theie nearest living relatives, *Aesculus glabra*, *Betula lutea*, and *Salix nigra*, only in the eastern United States, and five others, *Larix potaninii*, *Sorbus aucuparia*, *Picea polita*, *Mahonia lomariifolia*, and *Crataegus pinnatifida*, are in Japan and China. In both regions these species live adjacent to, or form a part of, conifer-hardwood forests characterized by species of *Abies*, *Pinus*, *Picea*, *Tsuga*, and other conifers, as well as numerous deciduous dicots. The fossil equivalents of the modern species now in the eastern parts of the northern continents were rare members of the conifer-hardwood forest at Eastgate, as judged from their low representation. They probably occupied sheltered sites where evaporation was reduced, thus compensating for the low summer rainfall in the basin.

Mixed-sclerophyll forest and chaparral are well developed in areas adjacent to mixed-conifer forest in the Sierra Nevada and northern Coast Ranges. They also adjoin similar forests in the Southwest and adjacent Mexico. The species that contribute to these sclerophyllous communities regularly occupy the warmer, drier borders of conifer forests at lower levels. For instance, in the southern part of the range of Sierra redwood, as in the Tule River basin near Camp Nelson, *Quercus chrysolepis* (cf. *Q. hannibali*) forms dense sclerophyll woodlands on rocky canyon walls adjacent to the valley forest dominated by *Sequoiadendron* (Plate 3, fig. 1). In this area there are associated shrubs--notably *Cercocarpus*, which forms pure stands--that have equivalents in the Eastgate flora. To the north, *Lithocarpus* occurs in the forest at North Grove, where it is represented by the shrub form *L. densiflora* var. *echinoides*. However, the tree occurs in the nearby Sierra at lower elevations, forming a dense sclerophyll forest on warmer slopes with *Arbutus menziesii* and *Quercus chrysolepis*, which are also regular understory associates in the lower part of the Sierran mixed-conifer forest. Judging from their present ecologic relations, and the abundant remains of their allied fossils in the flora, the warm, rocky walls of canyons in the Eastgate Hills at low levels close to the lake were probably covered with a dense sclerophyllous forest that contributed numerous leaves to the accumulating record.

There were additional members of this community at Eastgate whose modern counterparts do not now occur in the Sierra. Noteworthy in this group are 3 species whose nearest relatives are in insular southern California. Lyonothamnus parvifolius, a small-leaved extinct species, is allied to L. asplenifolius of the Channel Islands. The fossil Mahonia reticulata is most similar to M. insularis, a trailing 6-9 m (20 ft.) scrambler on Santa Cruz and Santa Rosa islands. And Cercocarpus ovatifolius resembles C. blancheae, which is on the Channel Islands and also on the higher, coastward slopes of the Santa Monica and Santa Ynez mountains, where climate is moist and very equable (M 65). In this connection, the tanbark oak-madrone (Lithocarpus-Arbutus) sclerophyll forest finds one of its southern outposts in the Santa Ynez Mountains above Santa Barbara, close to these insular species that are allied to the Eastgate fossils. Further, a relict Pinus ponderosa forest caps Figueroa Mountain about 24 km (15 mi.) north, at an elevation near 1,035 m (3,400 ft.). The general proximity of these plants that represent diverse vegetation zones in this area of highly equable climate is significant for interpreting the occurrence of their fossil relatives in the Eastgate and Middlegate floras.

Three fossil plants that appear to have contributed to sclerophyll vegetation at Eastgate are no longer represented by closely related species in California. Arbutus arizonica occurs from southern Arizona eastward to Texas and southward into Mexico, a region with ample summer rainfall. It is a regular member of the evergreen sclerophyll forest at the lower margin of the mixed-conifer forest. Cercocarpus breviflorus of New Mexico and the adjacent region, allied to C. eastgatensis, ranges up to the lower margin of the mixed-conifer forest and alternates with it on drier slopes, where it is a chaparral species. Robinia neomexicana, which is allied to R. californica, occurs with both of them throughout the southwestern United States. Finally, Eugenia is a component of the evergreen sclerophyll forest in both Mexico and southern China, where there is regular summer rainfall.

Table 6 groups the Eastgate species into the vegetation zones to which they most likely contributed. As suggested by the quantitative representation of Eastgate fossils, and the occurrences of living plants most similar to them, the area was covered chiefly by conifer-hardwood forest. The canopy was dominated by *Abies*, *Chamaecyparis*, *Larix*, *Picea*, *Pinus*, and *Sequoiadendron*, and the understory included species of *Acer*, *Aesculus*, *Amelanchier*, *Arbutus*, *Betula*, *Chrysolepis*, *Lithocarpus*, *Mahonia*, *Prunus*, and *Ribes*. Taxa whose descendants are presently in fir-subalpine forests, notably *Abies*, *Picea*, *Pinus*, *Tsuga*, and associated species of *Acer*, *Populus*, and *Sorbus*, also formed part of the conifer-hardwood forest. This resulted from the alleviation of severe drought stress by some summer rain and moderate summer temperature. Warm, south-facing slopes in the Eastgate Hills provided a favorable environment for patches of a sclerophyll forest of *Arbutus*, *Eugenia*, *Lithocarpus*, and *Quercus*. Some of the shrubs that contributed to it, such as *Cercocarpus*, *Heteromeles*, *Lyonothamnus*, *Mahonia*, and *Robinia*, probably formed dense, local communities on rocky walls of the scarp and especially on dry sites with thin soil provided by the hard welded tuffs, as did some of the shrubs of the nearby forest. Lake- and stream-margin sites were lined chiefly with deciduous trees and shrubs, such as *Acer*, *Amelanchier*, *Betula*, *Populus*, *Prunus*, *Salix*, and *Sorbus*, as well as some of the sclerophylls (*Lithocarpus*, *Quercus*) and the more mesic conifers (*Chamaecyparis*, *Sequoiadendron*).

TABLE 6

Eastgate Taxa Grouped According to the Vegetation Types they Probably Represented

FOSSIL SPECIES	AQUATIC	LAKE-BORDER & RIPARIAN	SCLEROPHYLL FOREST & *SHRUBLAND	CONIFER-HARDWOOD FOREST
Ceratophyllum praedemersum	x			
Nymphaeites nevadensis	x			
Sparganium nevadense	x			
Typha lesquereuxi	x			
Acer oregonianum		x	x	x
Acer tyrrelli		x		x
Amelanchier grayi		x	s	x
Betula thor		x		x
Fraxinus coulteri		x		x
Populus bonhamii		x		x
Populus cedrusensis		x		x
Populus eotremuloides		x		x
Populus payettensis		x		x
Populus pliotremuloides		x		x
Prunus chaneyi		x	x	x
Ribes stanfordianum		x		x
Salix desatoyana		x		x
Salix pelviga		x		x
Salix storeyana		x		x
Salix venosiuscula		x		x
Arbutus prexalapensis			X	x
Cercocarpus eastgatensis			s	x
Cercocarpus ovatifolius		s	s	x
Eugenia nevadensis				x
Heteromeles sonomensis			s	
Juniperus nevadensis			x	
Lyonothamnus parvifolius			s	x
Lithocarpus nevadensis			X	x
Mahonia reticulata		x	s	x

- cont. -

TABLE 6, cont.

FOSSIL SPECIES	AQUATIC	LAKE-BORDER & RIPARIAN	SCLEROPHYLL FOREST & *SHRUBLAND	CONIFER-HARDWOOD FOREST
Quercus hannibali			X	x
Quercus shrevoides			X	x
Quercus simulata			X	x
Rhamnus precalifornica			s	x
Robinia californica			s	x
Abies concoloroides				X
Abies laticarpus				X
Chamaecyparis cordillerae		X		X
Larix cassiana				X
Larix nevadensis				X
Picea lahontense				X
Picea sonomensis				X
Pinus alvordensis				X
Pinus balfouroides				X
Pinus sturgisii				X
Pseudotsuga sonomensis				X
Sequoiadendron chaneyi		x		X
Tsuga mertensioides				X
Acer nevadensis		x		x
Aesculus preglabra				x
Betula vera				x
Chrysolepis convexa				x
Crataegus newberryi		x		x
Mahonia macginitiei				x
Mahonia simplex				x
Sorbus idahoensis		x		x

*X = dominant

x = also in understory of lower conifer-hardwood forest

s = forms local shrub patches

CLIMATE AND ALTITUDE OF MIDDLEGATE BASIN

Climate

The Miocene climate of the Middlegate basin may be estimated from that now prevailing in areas with analogous vegetation. As outlined in the preceding pages, the two floras are generally similar but differ in that the Middlegate flora represents a dominant sclerophyll forest with conifer-hardwood forest taxa on nearby higher, cooler slopes, while while the Eastgate flora is that of a conifer-hardwood forest with patches of sclerophyll vegetation on nearby warmer slopes. Climatic conditions in the ecotone between these vegetation zones today thus provide a general indication of climate in the Miocene Middlegate basin.

Annual precipitation in the ecotone between mixed-conifer and sclerophyll forest generally totals about 890 mm (35 in.) or more at present. A similar total occurs in the eastern United States and eastern Asia in areas where plants allied to those in the floras live. The data suggest that at a _minimum_, annual precipitation was probably 760-890 mm (30-35 in.) in the Middlegate area, and not less than 890-1,015 mm (35-40 in.) at the Eastgate site. It no doubt increased on nearby higher slopes in both areas, and also in favorably oriented canyons. A moderate difference in precipitation, coupled with the diverse topographic settings, probably accounts for the contrasting representation of conifer-hardwood and sclerophyll forests in these floras.

There must have been some summer rainfall, because several genera in these floras live only in such areas. These include _Eugenia_ and _Robinia_ at Eastgate, and _Cedrela_, _Diospyros_, and _Hydrangea_ at Middlegate. In addition, a number of taxa in each flora have their nearest descendants only in areas with summer rain. These include species of _Aesculus_ (cf. _glabra_), Betula (cf. _papyrifera_ and _lutea_), _Mahonia_ (cf. _lomariifolia_), _Crataegus_ (cf. _pinnatifida_), and _Picea_ (cf. _polita_) that live either in the eastern United States or eastern Asia, or in the Rocky Mountains, notably

Acer (cf. grandidentatum), Populus (cf. angustifolia), Fraxinus (anomala), and Crataegus (cf. chrysocarpa), and Populus (cf. brandegeei) in Mexico.

To judge from the representation of comparable taxa in the nearby Miocene floras of western Nevada, summer rainfall was not evenly spread over the region. The amount was probably controlled chiefly by local topography and position with respect to west-facing broad topographic lows that connected with the coastal slope (see Axelrod, 1968, fig. 7). This may explain why the younger Pyramid flora (15 m.y.) has more numerous exotic members of the group than the Eastgate and Middlegate floras, even though the latter are older (18.0 m.y.). Furthermore, since the Middlegate basin was in the lee of a range to the west, less effective summer precipitation would be expected over the basin, and species with eastern alliances would be reduced. As noted earlier, the low ranges directly west of Middlegate have a basement of older Mesozoic rocks overlain by Miocene or Oligocene volcanic rocks. Since relatively recent faulting accounts for the present basin-range structure, there probably was a topographic high to the west, much of which has been removed by graben formation and erosion since the Miocene. Hence the low representation in the Middlegate basin floras of taxa that now occur in areas of summer rainfall becomes understandable, as does the younger aspect of these floras.

As for total warm-season rainfall in the Middlegate basin, the minimum may have been like that now at the western, drier limit of distribution of deciduous hardwood forest. In such areas, 50-75 mm (2-3 in.) occurs in June, July, and August, implying 150-225 mm (6-9 in.) as a warm-season minimum for the sheltered Middlegate basin. In the eastern Mediterranean basin, deciduous hardwood and conifer-hardwood forests in the mountains and the sclerophyll vegetation that adjoins them at lower levels are subject to a less severe summer drought than the comparable vegetation zones in California (Axelrod, 1980, fig. 15). In California, the summer drought period develops _gradually_ as rainfall progressively slackens in spring, and it is _gradually_ alleviated as precipitation slowly increases in the cooler months. In the eastern Mediterranean region, by contrast, relatively ample rainfall ceases _abruptly_ at the start of the drought period (June), which is _abruptly_ terminated by heavy rainfall in late September-October. As a result, the drought period is less pronounced in these modern forests that support _Carpinus_, _Castanea_, _Fagus_, _Liquidambar_, _Platanus_, _Tilia_, _Ulmus_, _Zelkova_, and other genera that occur also in the Miocene of Nevada. At that time much of central California was flooded by a sea that was as warm as, or little warmer than, the present Mediterranean. At this general latitude (37°-40°N), seasonal atmospheric changes in the Miocene may have led to a situation analogous to that in the Colchic region of southeastern Europe and adjacent Asia Minor, where sufficient moisture in the soil allows deciduous hardwoods to survive under temporary drought conditions in the interior that resulted from only about 6-9 in. precipitation during the summer season.

There was evidently ample summer rain in the Miocene of the western Mohave region, as shown by the Tehachapi flora 460 km (285 mi.) south (Axelrod, 1939). Similar relations are indicated by fossil floras to the north (Trout Creek, 49-Camp, Succor Creek). This suggests that the low Sierran barrier may have formed a rainshadow sufficient to reduce rainfall generally over west-central Nevada (Axelrod, 1957a). The composition of the fossil floras also indicates that the Middlegate area was transitional between a humid, temperate region to the north and a subtropical, subhumid one to the south. Clearly, this position between regions of regular cyclonic rainfall to the north and convectional precipitation to the south may also account for some of the uncertain precipitation in this transitional area. Nonetheless, the problem is compounded by the occurrence of numerous exotic taxa in the lower Fingerrock (16 m.y.) flora to the south and

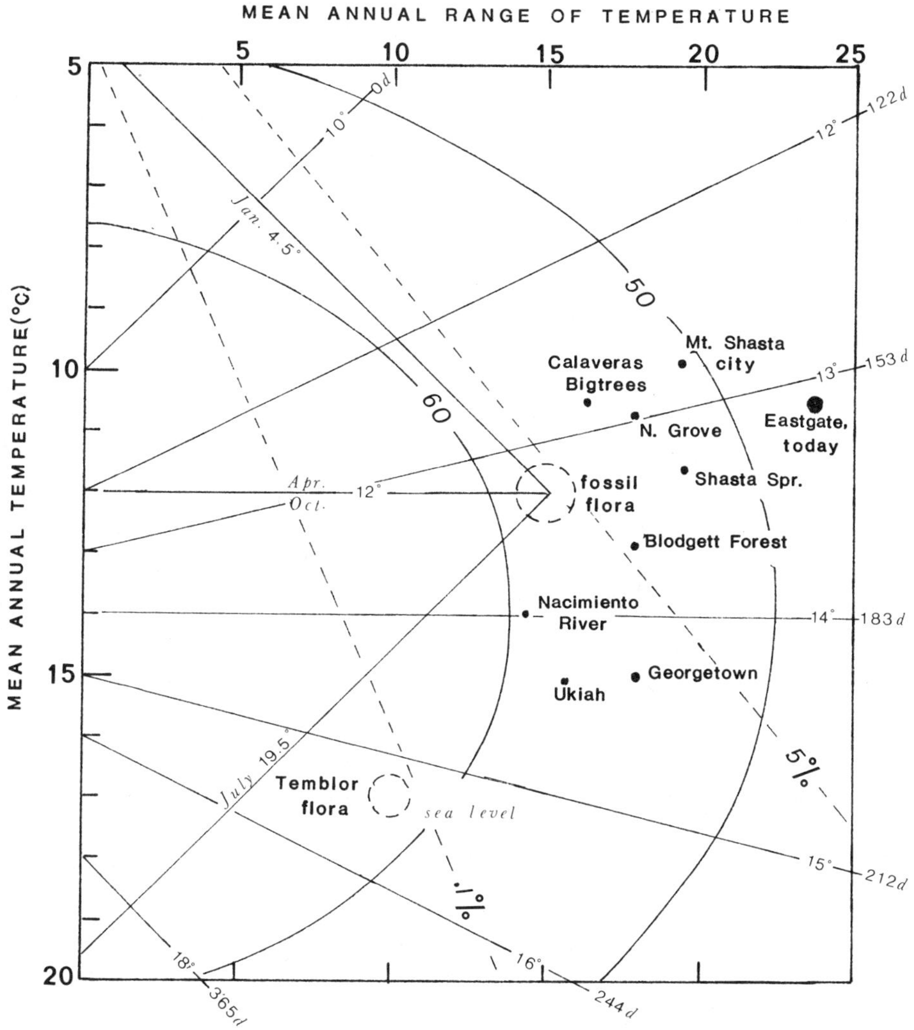

FIGURE 3. Thermal conditions suggested for the fossil floras. (Comparable data for the Temblor flora at sea level provide a basis for estimating altitude.)

the Pyramid flora (15 m.y.) to the west. However, these occurrences may reflect the position of these floras opposite topographic lows on the Sierran ridge that provided access to moisture-bearing storms. Coupled with the position of the Middlegate basin in the lee of a topographic high directly west, the rarity of exotic taxa in its floras seems understandable.

With respect to thermal conditions, present temperature in the zone between mixed-conifer and sclerophyll forest presumably approximates that in the fossil basin (fig. 3). Temperatures were moderate, to judge from the numerous descendant taxa that live today only on the western slope of the Sierra Nevada or in the Coast Ranges. Among these are _Abies_ (_bracteata_), _Chrysolepis_ (_chrysophylla_), _Chamaecyparis_, _Lithocarpus_, _Lyonothamnus_, _Mahonia_ (_insularis_), _Picea_ (_breweriana_), _Pseudotsuga_, _Quercus_ (_shrevei_), _Sequoiadendron_, and many others, as noted above. This means that the fossil floras have equivalent species in areas where the mean monthly range of temperature averages 16-18°C (in the Sierra Nevada) to 14-15°C (in the Coast Ranges). This is 5-8°C lower than that now in western Nevada, as charted in fig. 3.

That the range of temperature in the Middlegate basin probably was less than that now in the Sierra Nevada is indicated by several lines of evidence. In the first place, the ocean was warmer, so the range of temperature on land was less than at present. Second, the Sierra Nevada was not yet elevated appreciably (Axelrod, 1957a, 1980), so moderating effects penetrated into the interior to keep winters mild. Third, the abundance of laurels in the Miocene Carson Pass and Ebbetts Pass floras from the Sierran crestal area indicates mild winter climate only 120 miles to the west. Fourth, the well-developed sclerophyll forest implies little snow, because heavy snow accumulation physically breaks the trees, as is evident from occasional heavy snowfalls in the Coast Ranges today (Axelrod, 1976a). And fifth, the polar ice-cap had not yet developed, so frigid winter air masses were not in existence at this latitude. For these reasons, the range of temperature was probably more nearly like that now in the inner Coast Ranges rather than in the Sierra Nevada, where relict stands of sclerophyll forest are confined chiefly to steep-walled, protected canyons or to flats where they are often under the protective cover of tall conifers.

A mean monthly range of temperature (A) near 15°C, and a mean annual temperature (T) of about 12°C, is thus inferred for the flora, as depicted in figures 3 and 4. This gives mean temperatures of 4.5°C in January and about 19.5°C in July, with mean April and October temperatures of approximately 12°C. (For methodology, see Bailey, 1960, 1964; also Axelrod and Bailey, 1976, and Axelrod, 1981.) The data imply that the warmth of climate was W 13.3°C, or that 162 days had temperatures warmer than this, which represents a measure of the growing season.

These figures, which are an average for the basin, can be adjusted for the different sites. A slightly cooler climate, but still within the range of variation shown in figures 3 and 4, is indicated for Eastgate. This would give a slightly higher equability rating to the Eastgate area, say M 58 as compared with M 57 at Middlegate. Eastgate had a more effective precipitation, owing to its position on the windward side of the lake, whereas the Middlegate site was not only on a south slope but in the lee of a range which has since been sundered by faulting.

An estimated frost frequency of about 5% of annual hours (fig. 3) raises the question of the probable amount of snow during the cool season. Although heavy snowfall typifies the western mixed-conifer forests today, it seems in view of the medium equability rating (M 57) estimated for the area that snows could not have been very heavy or long in duration. This is consistent with estimates that Miocene sea-surface temperatures at this latitude were warm temperate, as judged from the critical data presented by Hall (1964). As he and

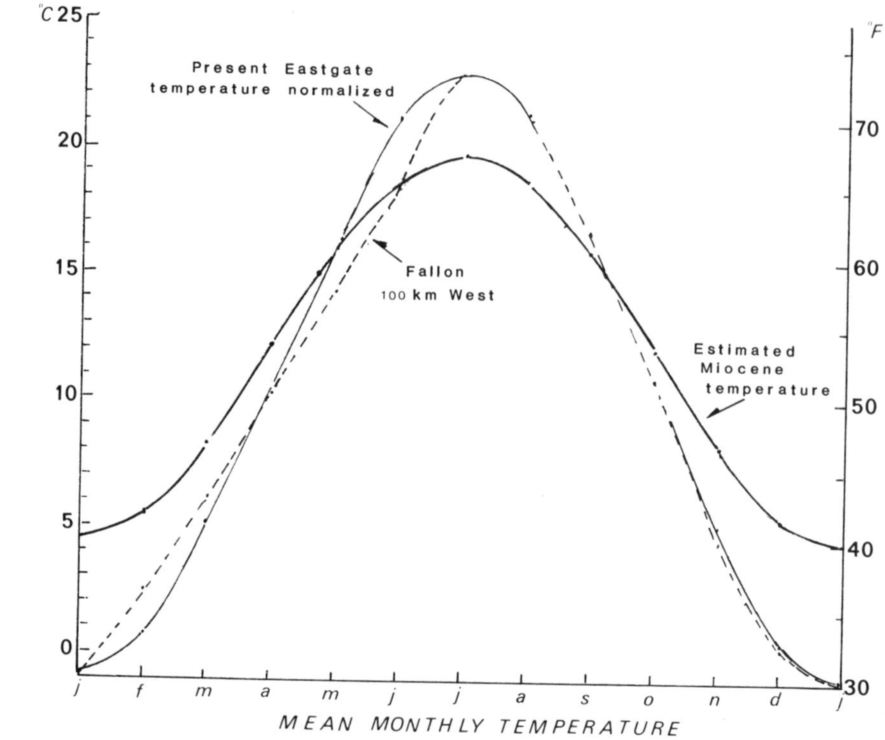

FIGURE 4. Inferred mean monthly temperatures for the fossil floras, compared with those in western Nevada today.

others (e.g., Hedgpeth, 1957) emphasize, it is the <u>duration</u> (i.e., warmth) of shallow-water temperatures required for their reproduction and establishment that determines distribution of marine mollusks today. Estimates of these temperatures obviously give a more reliable measure of marine paleoclimate than data based solely on minimum winter temperature or on mean annual temperature alone (e.g., Durham, 1950). Hence, Hall's estimates of shallow-water paleotemperature are in much closer agreement with the data suggested by Miocene floras bordering the coast (Temblor, Topanga, Puente, Carmel). Furthermore, since large ice-caps were not yet in existence at high latitudes, winters must have been relatively mild. This is also implied by the nature of middle

Miocene floras in the Sierra Nevada, where abundant laurels and other frost-sensitive taxa are recorded. The evidence suggests that although the Middlegate basin probably had regular light dustings of snow each winter, conditions sufficient to bring deep drifts or blizzards were not yet in existence.

Altitude

In earlier papers (e.g., Axelrod, 1956, 1957a, 1965, 1968), the general altitude of the basins over western Nevada during Miocene time was estimated to be near 600-750 m (2,000-2,500 ft.). This was inferred from the general climatic indications of the few floras then known from the region, as compared with those in the Sierra Nevada and coastal California. The differences in climate that they indicated seemed compatible with only a moderate rainshadow over the area east of the Sierran axis, and hence a low divide in its central and northern sector during the later Miocene.

Altitude may be estimated from the mean annual temperature (*T*) inferred for two fossil floras of generally similar age at about the same latitude, one at sea level, the other in the interior. The difference in *T* estimated for the floras, multiplied by normal terrestrial lapse rate (1°C/183 m), approximates the difference in altitude between them (Axelrod, 1964, 1968, 1981; Axelrod and Bailey, 1976). However, the estimate *T* must be based on both the warmth and equability ratings inferred for the fossil floras. This is necessary because *T*, when considered alone, has no close climatic significance. This is shown quite clearly by thermal data for Harar (Ethiopia), Lima (Peru), and Mosul (Iraq), which all have a mean temperature of 20°C (68°F) but encompass tropical, warm temperate, and temperate climates, respectively. In the same way, Rio Grande (Argentina), Dutch Island (Alaska), Nemuro (Japan), and Igriz (Russia) all have a *T* of about 4.4°C (40°F), but occur in tundra and taiga, as well as cool temperate climate that supports conifer-hardwood forest. Furthermore, stations with a T near -6.7°C (20°F) may indicate either a full glacial climate (El Misti, Peru) or a cold temperate climate suited to taiga forest (Ft. Yukon, Alaska). Obviously, previous estimates of paleotemperature based on mean annual temperature alone have little meaning in terms of the climates in which ancient forests lived (cf. Frakes and Kemp, 1972, 1973; Wolfe and Hopkins, 1967).

As shown in figure 3, the Eastgate flora is estimated to have had a T of 12°C and a mean annual range (A) of 15°C, which gives a warmth of about W 13.3°C and an equability of M 57-58. The Temblor flora of somewhat younger age (15 m.y.) in the coastal strip is preserved in the basal beds of the Temblor Formation north of Coalinga, California, which is situated slightly south of the latitude of the Middlegate basin. The Temblor flora includes species of Glyptostrobus, Keteleeria, Pinus, Arbutus, Carya, Castanea, Cedrela, Cornus, Lithocarpus, Magnolia, Mahonia, Nyssa, Persa, Quercus (5 spp.), Robinia, and Zelkova (Renney, 1972). Their thermal relations imply a T of about 16°C (62°F) and an A near 10°C. Since the flora is somewhat younger than the Middlegate and Eastgate floras, a T of 17°C is estimated for the coastal area in the middle Miocene (18.0 m.y.). This implies a difference of about 5°C between the coastal strip (T 17°C) and that inferred for the Nevada floras (T 12°C). With a normal terrestrial lapse rate (183 m/-1°C; 333 ft./-1°F), the lakeshore of the Middlegate basin is estimated to have been near 915 m (3,000 ft.).

From these data, the minimum altitude necessary to support fir-subalpine conifer forest in the region can be estimated. The lower margin of this forest (W 11.5°C) has only 105 days with mean temperatures warmer than 11.5°C (53°F). As shown in figure 3, about 5°C in T separates the thermal level indicated for the Eastgate flora (T 12°C) and the lower margin of fir-subalpine forest. Using a normal terrestrial lapse rate (1°C=183 m; 1°F=333 ft.), relief would have to be near 915 m (3,000 ft.) above the lake for a thermal level sufficiently low to support fir-subalpine conifer forest. From our estimates (lake level, 915 m a.s.l.; subalpine zone 915 m above the lake) the lower margin of fir-subalpine forest had an elevation of about 1,830 m (6,000 ft.) above sea level in this region in the middle Miocene. Since cold pockets would be present on northeast-facing slopes, the forest no doubt descended to somewhat lower levels in favorable sites.

The estimate that the lower margin of fir-subalpine forest was probably near 1,830 m in the adjacent region may at first seem startling, because it is now close to 2,735 m (9,000 ft.) in the mountains to the east. However, fir-subalpine forest is at 1,825 m (6,000 ft.) on the west slope of the Sierra Nevada at Sacramento Camp on California State Highway 50, and near 1,670 m (5,500 ft.) two miles east of Emigrant Gap on State Highway 40. Furthermore, it is as low as 1,300 m (4,300 ft.) in the Siskiyou Mountains of northwestern California where there is a sharp ecotone between mixed-conifer forest and the fir-subalpine conifer forest dominated by *Abies magnifica*, *Picea breweriana*, *Pinus monticola* and *Tsuga mertensiana*. Clearly, with a low Sierran divide to the west, ample summer rainfall, a normal lapse rate, and a lower range of mean monthly temperature, conditions were suitable for fir-subalpine forest at much lower levels in the middle Miocene than in western Nevada today. While an estimate of the lower margin of the Miocene fir-subalpine forest at 1,830 m (6,000 ft.) seems consistent with the paleoecology suggested by the floras of the Middlegate basin, the forest may have been 150 m (500 ft.) lower, or near 1,670 m (5,500 ft.). This would be consistent with the numerous sclerophylls in the floras which have close allies in the central Coast Ranges of California.

COMPARISON OF THE FLORAS

Terrain

To appreciate the similarities and differences between the Middlegate and Eastgate floras, it is appropriate to recall that they had contrasing topographic settings. Terrain near the Middlegate site was relatively low, as judged from the fine-grained sedimentary rocks that lap onto the older volcanic terrain. Since higher hills rose to the north, the site received plant remains from vegetation that mostly had warmer and drier southerly exposures. This was accentuated by its position in the lee of the south end of the Miocene Clan Alpine Range. On the other hand, the Eastgate flora lived on the windward, western front of the low hills at the site of the present Eastgate Hills, which were bounded by an active fault scarp. Active uplift indicates that a narrow shore zone fronted the scarp, and that canyons were relatively deep and narrow, with a youthful "V" profile. Since the canyons drained generally west, they provided north-facing, sheltered, cooler slopes near the lakeshore. Rocky dry sites with shallow soil were on the spurs along the scarp and on the south-facing canyon walls in the hard welded tuffs. Current directions in the thick, pebbly sandstone directly under the fossil flora imply that sediment transport was from the south to southeast. Thus much of the plant debris was apparently transported from a forest that reached down to the basin on cooler, north-facing slopes south of the fossil locality.

Composition

Although the differences in topographic setting and local climates on opposite sides of Middlegate Lake account for the different representation of taxa, and of vegetation to which they contributed, there are nonetheless important similarities between the floras. Of the total 80 species in the combined floras, 39 (50%) are common to both, as listed below. Of these, 24 are trees, 13 are shrubs, and 2 are herbaceous aquatics. At Eastgate, the species common to both floras make up 85% of the total sample, while at Middlegate, which is dominated by one species of oak (85%), 96% of the specimens represent species that occur in both floras. These very high percentages indicate that the habitats near the sites of plant accumulation were similar.

Species Common to the Eastgate and Middlegate Floras

Pinaceae
- Abies laticarpus
- Picea lahontense
- Picea sonomensis
- Pinus sturgisii
- Pseudotsuga sonomensis
- Tsuga mertensioides

Cupressaceae
- Chamaecyparis cordillerae
- Juniperus nevadensis

Taxodiaceae
- Sequoiadendron chaneyi

Typhaceae
- Typha lesquereuxi

Salicaceae
- Populus bonhamii
- Populus cedrusensis
- Populus eotremuloides
- Populus payettensis
- Populus pliotremuloides
- Salix pelviga
- Salix storeyana
- Salix venosiuscula

Betulaceae
- Betula thor
- Betula vera

Fagaceae
- Chrysolepis convexa
- Lithocarpus nevadensis
- Quercus hannibali
- Quercus shrevoides
- Quercus simulata

Berberidaceae
- Mahonia macginitiei
- Mahonia reticulata
- Mahonia simplex

Nymphaeaceae
- Nymphaeites nevadensis

Rosaceae
- Cercocarpus eastgatensis
- Heteromeles sonomensis
- Lyonothamnus parvifolius
- Sorbus idahoensis

Leguminosae
- Robinia californica

Aceraceae
- Acer nevadensis
- Acer oregonianum
- Acer tyrrelli

Ericaceae
- Arbutus prexalapensis

Oleaceae
- Fraxinus coulteri

As Table 7 shows, the Eastgate flora contains 7 trees, 7 shrubs, and 2 aquatics apparently unique to it, while the Middlegate flora has 14 trees, 10 shrubs, and 1 aquatic that were not collected at the Eastgate site. The Eastgate species not recorded at Middlegate are mainly forest taxa that required moister, cooler environments, notably 5 conifers, *Abies concoloroides* (cf. *concolor*), *Larix cassiana* (cf. *potaninii*), *L. nevadensis* (cf. *occidentalis*), *Pinus alvordensis* (cf. *murrayana*), and *P. balfouroides* (cf. *balfouriana*), as well as *Aesculus preglabra* (cf. *glabra*) and *Eugenia*.

TABLE 7

Species Restricted to Either the Middlegate or the Eastgate Flora

FOSSIL SPECIES	MIDDLEGATE FLORA	EASTGATE FLORA
Equisetaceae		
Equisetum alexanderi	x	
Pinaceae		
Abies concoloroides		x
Abies scherri	x	
Larix cassiana		x
Larix nevadensis		x
Pinus alvordensis		x
Pinus balfouroides		x
Sparganiaceae		
Sparganium nevadense		x
Salicaceae		
Salix desatoyana		x
Salix owyheeana	x	
Salix wildcatensis	x	
Juglandaceae		
Juglans nevadensis	x	
Betulaceae		
Alnus harneyana	x	
Alnus largei	x	
Fagaceae		
Chrysolepis sonomensis	x	
Berberidaceae		
Mahonia trainii	x	
Ceratophyllaceae		
Ceratophyllum predemersum		x
Platanaceae		
Platanus dissecta	x	
Platanus paucidentata	x	
Hydrangeaceae		
Hydrangea ovatifolius	x	
Grossulariaceae		
Ribes stanfordianum		x
Rosaceae		
Amelanchier grayi		x
Cercocarpus antiquus	x	
Cercocarpus ovatifolius		x

- cont. -

TABLE 7, cont.

FOSSIL SPECIES	MIDDLEGATE FLORA	EASTGATE FLORA
Crataegus middlegatensis	x	
Crataegus newberryi		x
Crataegus pacifica	x	
Prunus chaneyi		x
Prunus moragensis	x	
Leguminosae		
Gymnocladus dayana	x	
Aceraceae		
Acer middlegatensis	x	
Acer negundoides	x	
Acer scottiae	x	
Meliaceae		
Cedrela trainii	x	
Hippocastanaceae		
Aesculus preglabra		x
Rhamnaceae		
Ceanothus precuneatus	x	
Rhamnus precalifornica		x
Myrtaceae		
Eugenia nevadensis		x
Stryacaceae		
Stryax middlegatensis	x	
Ebenaceae		
Diospyros oregonianum	x	
Oleaceae		
Fraxinus millsiana	x	
Totals	25	16

Furthermore, there are different species of Cercocarpus, Crataegus, Prunus, and Salix in each flora, indicating minor ecologic preferences, with most of those at Middlegate less mesic in their requirements. Although both floras represent vegetation near the ecotone between sclerophyll and conifer-hardwood forest, the Eastgate site, with nearby sheltered, cool, moist canyons, has more species from mesic and cooler habitats. The Middlegate flora occupied a sunnier, drier area where the conifer-hardwood forest was restricted to moister upland sites sufficiently removed from the lake to contribute fewer structures to the accumulating record. As a result, such subhumid species as Abies scherri (cf. bracteata), Ceanothus precuneatus (cf. cuneatus), Crataegus middlegatensis (cf. chrysocarpa), Fraxinus millsiana (anomala), Juglans nevadensis (californica), Platanus paucidentata (cf. racemosa), and Styrax middlegatensis (cf. californica) appear at Middlegate, but not at Eastgate. Their absence (or at least rarity) at Eastgate is understandable if that area was dominated chiefly by forest species that required more abundant rainfall and somewhat lower temperature.

Vegetation

Table 8, which indicates the number of species contributing to the different vegetation types in the floras of the Middlegate basin, suggests that they are very similar. However, the critical point is that their representation differs because their topographic-climate settings were dissimilar. The fundamental differences between the floras are shown by the data which indicate the relative abundance of specimens representing each taxon (Tables 2 and 4). As emphasized earlier, the counts of species abundance do not necessarily indicate dominance of taxa in the vegetation zones, as some investigators have assumed; rather, they provide evidence of relative proximity to the site. The presence of a few leafy twigs of Sequoiadendron and other rare forest conifers in the Middlegate flora, which is dominated by sclerophyll forest taxa, implies that Sierra redwood and its forest associates lived farther upstream and on more distant, moister slopes. By contrast, the occurrence at Eastgate of large branchlets and cones of Sequoiadendron indicates that this and other conifers, which are also relatively abundant, lived much closer to that area of plant accumulation. The quantitative data which indicate that conifer-hardwood forest dominated the Eastgate area, and sclerophyll forest the Middlegate site, are consistent with the physical setting we have adduced for the floras.

TABLE 8

Representation of Vegetation Types in Middlegate and Eastgate Floras*

VEGETATION	MIDDLEGATE SPECIES	EASTGATE SPECIES
Sclerophyll forest & shrubland	30 48	16 49
Conifer-hardwood forest	48	49
Riparian/lakeshore woodland	25	23
Aquatic	3	4

*Numbers indicate total species in each vegetation zone. Note that some taxa occurred in more than one zone. Data from Tables 3 & 6.

Some taxa might be placed in either conifer-hardwood forest or sclerophyll forest, for their modern allies occur in both. For example, the modern equivalents of _Lithocarpus_ and _Quercus_ (_hannibali_), _L. densiflorus_ and _Q. chrysolepis_, contribute to sclerophyll forest (Plate 2, figs. 1 and 2) and also form an understory in the lower part of the mixed-conifer forest in the Sierra Nevada. Judging from the representation of species in the Middlegate flora, these taxa probably formed the dominants of the sclerophyll forest that bordered the basin there, ranging upslope to contribute also to the lower part of conifer-hardwood forest. But in the Eastgate flora, these abundant taxa were probably members of the conifer-hardwood forest that lived close to the lake, judging from the dominance of conifer taxa in our sample. Nonetheless, in that area these species also entered into the composition of the sclerophyll forest, which probably had a more local occurrence in the lowlands there.

Of course, the taxa can be grouped in a manner that implies a different representation of vegetation in the region. For example, at Middlegate the riparian species of _Acer_, _Populus_, and _Salix_ could be considered as part of sclerophyll forest, together with riparian taxa that occur in both mixed-conifer and sclerophyll forest, notably _Acer_,

Fraxinus, Platanus, Populus, and Salix. On this basis, sclerophyll forest with associated riparian vegetation would rise to 95%. This only tends to emphasize what is already apparent--that sclerophyll vegetation dominated near the site of deposition to the virtual exclusion of conifer-hardwood forest. Similarly, if *Quercus hannibali* (cf. *chrysolepis*) is considered typical of sclerophyll vegetation at Eastgate, that community makes up 55% of the material. However, if it is regarded as a major forest component, then mixed-conifer forest taxa increase to 86%; and if we add the stream- and lake-border trees and shrubs to the forest category, then 88% of the Eastgate collection is represented by taxa of mixed-conifer forests, which seems consistent with the nature of the sample. To appreciate the differences in the representation of vegetation (and taxa) on opposite shores of the lake basin, we turn to a further analysis of the communities in each area.

Sclerophyll Forest

Table 9 shows that sclerophyll forest has a higher numerical representation at Middlegate. This is expectable, since it occupied warmer, south-facing slopes, whereas such sites were more restricted at Eastgate. To judge from the ecologic requirements of *Q. chrysolepis* today, its fossil equivalent, *Q. hannibali*, probably dominated the steep, rocky walls of canyons bordering stream margins, descending to the lakeshore where it was mixed with riparian and lake-border trees in well-drained sites. This may account in part for its high representation in both floras. Another important factor is the large size and spreading habit of the oak, as well as the heavy texture of its durable leaves, which entered the record in large numbers.

Most of the other members of the sclerophyll forest are less common at Eastgate. This presumably reflects the cooler, moister climate there, as judged from the high representation of conifers. The single specimen of *Eugenia* at Eastgate implies that it was a rare tree there, and if

TABLE 9

Taxa Typical of Sclerophyll Forest in Middlegate and Eastgate Floras

DIAGNOSTIC TAXA	MIDDLEGATE SPECIMENS	EASTGATE SPECIMENS
Abies scherri	2	
Arbutus prexalapensis	14	7
Cedrela trainii	3	
Cercocarpus antiquus	141	
Cercocarpus eastgatensis	7	28
Cercocarpus ovatifolius		104
Eugenia nevadensis		1
Heteromeles sonomensis	2	4
Juniperus nevadensis	7	
Lithocarpus nevadensis	126	(1,497)
Lyonothamnus parvifolius	13	9
Platanus paucidentata	2	
Quercus hannibali	5,766	(2,568)
Quercus shrevoides	252	7
Quercus simulata	6	8
Taxa	13	8(2)
Specimens	6,341	168
OTHER TAXA*		
Acer oregonianum	x	x
Acer tyrrelli	x	x
Amelanchier grayi		x
Crataegus middlegatensis	x	
Mahonia macginitiei	x	x
Mahonia reticulata	x	x
Mahonia simplex	x	x
Platanus dissecta	x	
Prunus chaneyi		x
Prunus moragensis	x	x
Ribes stanfordianum		x
Robinia californica	x	x
Taxa	9	10

*These are chiefly riparian or moist-valley species that occur also in conifer-hardwood forests (Table 11).

( ) Most of these taxa contributed to conifer-hardwood forest at Eastgate and are counted there.

present at Middlegate, it probably was confined to moister upland sites.

Other trees no doubt entered into the composition of the sclerophyll forest, much as they do today (see Cooper, 1922). On its more mesic borders close to conifer-hardwood forest, some deciduous hardwoods, notably species of *Acer*, *Betula*, and *Platanus*, as well as various shrubs--*Amelanchier*, *Crataegus*, *Mahonia*, *Prunus*, *Ribes*--probably entered the sclerophyll forest. This sclerophyll forest was richer than its surviving counterpart, for it included taxa whose descendants are now found only in areas of ample summer rain, taxa that disappeared from California and border areas during the Pliocene and early Pleistocene.

Shrubland

This term is preferred to *chaparral* simply because chaparral as it is developed in California and Arizona today probably was not present in the area. The taxa that contributed to shrubland (Table 10) included some sclerophylls, but they were not common, and hard-leaved taxa like scrub oaks or members of *Ceanothus* subgenus Cerastes evidently were rare in the region.

TABLE 10

Taxa Typical of Shrubland in Middlegate and Eastgate Floras

TAXA		MIDDLEGATE SPECIMENS	EASTGATE SPECIMENS
Ceanothus precuneatus		1	
Cercocarpus antiquus		141	
Cercocarpus eastgatensis		7	28
Cercocapus ovatifolius			104
Fraxinus millsiana		5	
Heteromeles sonomensis		2	4
Rhamnus precalifornica			1
Robinia californica		13	2
Styrax middlegatensis		1	
	Taxa	7	5
	Specimens	170	139

Shrubland occupied drier, rocky sites where soil was thin. The community was evidently more diverse at Middlegate, consistent with the better development of warmer and drier sites in that area. The high representation of *Cercocarpus* in these floras probably reflects local dominance near the lake. This is consistent with the nature of shrub communities in areas adjacent to forests today, where one or two species may dominate. Such a relationship can be seen in the southern Sierra Nevada bordering *Sequoiadendron* groves, in the Tule River basin near Camp Nelson, and on the South Fork of the Kaweah River near the Garfield Grove. There, pure stands of *Cercocarpus betuloides* and *Quercus chrysolepis* cover exposed, rocky slopes close to *Sequoiadendron* and nearly in the shade of the mixed-conifer forest. In these areas they are adjacent to a river and contribute to the debris that it receives. In this way a large patch or patches of pure or nearly pure *Cercocarpus* could contribute a disproportionate number of specimens to the accumulating record. Thus the high representation of *Cercocarpus* specimens at these localities may be misleading as an indicator of a landscape dominated by shrubland. Furthermore, the shrubland probably was seral in nature, much as it is today at or near the forest margin.

Shrubs from the conifer-hardwood forest probably contributed to shrubland at both the Middlegate and Eastgate areas, much as they do today in forested regions. At Middlegate these probably included *Crataegus* (cf. *chrysocarpa*), *Prunus* (cf. *emarginata*), and *Mahonia* (cf. *pinnata*), whereas at Eastgate the associates probably were *Amelanchier* (cf. *florida*), *Prunus* (cf. *demissa*) and *Ribes* (cf. *nevadense*), as well as *Mahonia* (cf. *insularis*). The shrubland evidently was a mixed community of both deciduous and evergreen taxa in this ecotonal region between sclerophyll forest and conifer-hardwood forest. Similar shrublands that occur today in the Rocky Mountains were captioned "Petran chaparral" by Clements (1920).

Conifer-Hardwood Forest

Table 11 shows that conifer-hardwood forest formed an important community close to the lakeshore at the Eastgate site. The 12 conifers that contributed to the canopy of the forest account for 820 specimens. By contrast, the 7 conifers in the Middlegate forest are represented by only 73

TABLE 11

Taxa Typical of Conifer-Hardwood Forest in Middlegate and Eastgate Floras

TAXA	MIDDLEGATE SPECIMENS	EASTGATE SPECIMENS
TREES-Canopy		
Abies concoloroides		11
Abies laticarpus		14
Chamaecyparis cordillerae	4	29
Larix nevadensis		7
Picea lahontense	18	224
Picea sonomensis	24	85
Pinus alvordensis		22
Pinus balfouroides		4
Pinus sturgisii	9	52
Pseudotsuga sonomensis	3	57
Sequoiadendron chaneyi	12	277
Tsuga mertensioides	3	38
TREES-Subcanopy		
Acer middlegatensis	16	
Acer negundoides	33	
Acer oregonianum	86	2
Acer tyrrelli	28	90
Aesculus preglabra		7
Betula thor	14	250
Betula vera	7	2
Cedrela trainii	3	
Diospyros oregonianum	1	
Eugenia nevadensis		1
Lithocarpus nevadensis	(126)	1,497*
Quercus hannibali	(5,766)	2,568*
Quercus simulata	6	8
SHRUB LAYER		
Acer nevadensis	4	2
Amelanchier grayi		8
Crataegus middlegatensis	4	
Crataegus newberryi		1
Crataegus pacifica	1	

- cont. -

TABLE 11, cont.

TAXA	MIDDLEGATE SPECIMENS	EASTGATE SPECIMENS
Heteromeles sonomensis	2	4
Hydrangea ovatifolius	2	
Mahonia macginitiei	1	
Mahonia reticulata	5	52
Mahonia simplex	2	12
Mahonia trainii	1	
Prunus chaneyi		3
Prunus moragensis	1	
Ribes stanfordianum		3
Robinia californica	13	2
Sorbus idahoensis	2	22
Taxa	28(2)	31
Specimens	305	5,354

*Some of these contributed to sclerophyll forest at Middlegate.

( ) Most of these contributed to sclerophyll forest at Middlegate.

specimens, of which the small, light-winged seeds of spruces make up more than half (42 specimens) the total. These probably were transported to the area from higher hills to the north; this would be expected during autumn, when numerous seeds were being shed. The associated trees and shrubs that formed the forest subcanopy and shrub layers are also more abundantly represented by specimens at Eastgate, though the number of taxa is comparable, with 19 species at Eastgate and 21 at Middlegate. It is apparent that the forests in each area are broadly similar and that the different representation of their taxa reflects proximity to the lakeshore and to the site of plant accumulation. In this connection, it is noteworthy that 3 of the trees that are well represented at Middlegate, _Acer negundoides_, _A. tyrrelli_, and _Betula thor_, are typical of riparian-border sites. By eliminating them from consideration, the representation of Middlegate specimens that contributed to conifer-hardwood forest would be reduced materially.

The high representation of conifers at Eastgate has been noted, and the occurrence of cones and large branchlets of *Sequoiadendron* emphasized. This is understandable on the assumption that favorable terrain near at hand enabled conifer forest taxa to extend closer to the lakeshore than in the Middlegate area, where the forest was confined to higher, more distant, and moister sites. This is consistent with the representation in these floras of taxa that indicate summer rainfall, for the species at Eastgate indicate more mesic conditions than do those at Middlegate. Whereas comparatively mesic species of *Aesculus*, *Crataegus*, and *Eugenia* are at Eastgate, the Middlegate flora has *Cedrela*, *Diospyros*, and *Hydrangea*, which are relatively warmer in their requirements, consistent with the local setting.

Depending on the site, the conifer-hardwood forest no doubt varied in composition locally, much as it does today. Valley bottoms and cooler slopes at lower levels probably were dominated chiefly by *Sequoiadendron*, *Chamaecyparis*, *Abies*, *Larix*, and *Pseudotsuga*, whereas *Pinus* was confined primarily to the better-drained, somewhat drier, and warmer slopes and ridges. Intermediate sites probably supported a rich mixed forest. Similarly, riparian trees not only lined watercourses in the forest, but also occupied moist swales and seepages. In this connection, *Chamaecyparis* regularly descends along stream margins down canyons with cold-air drainage to lower levels, as may be seen today on the Sacramento River west, and also south of the town of Mount Shasta. In the latter area, near Castle Crags State Park, it occurs in sclerophyll forest dominated by *Quercus chrysolepis*. Similarly, in the southern part of its area *Sequoiadendron* descends into sclerophyll vegetation along rivers and in canyons with cold-air drainage (Plate 3, fig. 1). This suggests that at Eastgate, conifers reached the lakeshore area, where they locally mingled with evergreen oaks as well as deciduous members of the lakeshore woodland.

Fir-Subalpine Conifer Forest

Judging from the requirements of their modern relatives, several taxa in the Eastgate and Middlegate floras probably reached up to form a fir-subalpine conifer forest in areas of suitable elevation in the bordering region. As noted above, such hills would have been at least 900 m (3,000 ft.) above the lake level, which stood close to 900 m (3,000 ft.). From such sites, these trees no doubt reached down to lower levels on cooler, northeast-facing slopes, from which their structures were carried to lower elevations. Even though they are represented primarily by the light, winged seeds of fir, spruce, hemlock, and pine, this indicates their occurrence in the Middlegate basin. This applies also to the rare remains of *Acer* (cf. *glabrum*), *Populus* (cf. *tremuloides*), and *Sorbus* that probably contributed also to fir-subalpine forests, much as their descendants do today. Transport from distant mountains many miles away seems unlikely. As noted above, and discussed elsewhere (Axelrod, 1976a), most of these taxa whose modern correlatives are regular members of fir-subalpine conifer forests contributed also to conifer-hardwood forest during the Neogene and earlier. At that time there was regular summer rain, temperatures were mild, evaporation was reduced, and hence seedlings could become established in forests below the fir-subalpine zone. Such occurrences may be seen today in favorable sites in the Klamath-Siskiyou Mountains of northwestern California, and in areas farther north where summer rain gradually increases and lower summer temperatures reduce drought stress.

Riparian and Lakeshore Woodland

The relatively lower terrain at Middlegate probably favored a more optimum development of riparian and lake-border woodland. The broader river valleys and their flood-plains would support dense woodlands with more numerous trees than would have lived in the narrow canyons we have inferred for the Eastgate site. Furthermore, the lakeshore woodland would also have a better development at Middlegate because there was a wider shore area, whereas Eastgate was backed by a steep, active scarp.

These inferences seem consistent with the high representation (85%) of Q. hannibali at Middlegate, where it presumably formed dense stands overhanging the stream and lake borders. As for the less common species (Table 12), several are allied to plants that attain optimum development in areas of warmer climate, as in the woodland belt which has a richer representation at Middlegate. Furthermore, of the species represented by a single specimen at Eastgate, Fraxinus, which is typical of woodland vegetation, has a better representation at Middlegate. Conversely, we would expect that birch would be more abundant at Eastgate, for it regularly occurs with conifer-hardwood forest, which is more richly represented there. Species of Populus and Salix that are

TABLE 12

Taxa Typical of Riparian and Lakeshore Woodland in Middlegate and Eastgate Floras

TAXA		MIDDLEGATE SPECIMENS	EASTGATE SPECIMENS
Populus bonhamii		40	4
Populus cedrusensis		1	24
Populus eotremuloides		7	15
Populus payettensis		6	6
Populus pliotremuloides		2	10
Salix desatoyana			2
Salix pelviga		6	48
Salix storeyana		33	5
Salix venosiuscula		1	
Salix wildcatensis		1	
Alnus harneyana		1	
Betula thor		14	250
Platanus dissecta		1	
Platanus paucidentata		2	
Acer negundoides		33	
Fraxinus coulteri		5	1
	Taxa	15	10
	Specimens	153	365

more abundant at Eastgate are, in general, those whose modern correlatives require or prefer cooler sites. The data thus support the suggestion that the differences in taxa are due chiefly to local environmental setting. We realize that some of the differences reflect the factors of transport and preservation, but these are difficult to assess in most cases.

AGE

The Eastgate and Middlegate floras have the same stratigraphic position, in the upper 10 m of the Middlegate Formation, and are therefore of the same age. The Middlegate flora was earlier thought to be late Miocene or Pliocene, then considered 10-12 m.y. old. The reasons for that assignment were (*a*) the low representation of taxa whose nearest descendants are in areas with ample summer rainfall, as in the eastern United States or eastern Asia, and (*b*) the occurrence in the overlying Monarch Mill Formation of scrappy mammalian remains (chiefly broken limb bones) that appeared to be possibly of Hemphillian age (see Axelrod, 1956, p. 204).

Present evidence now shows that members of the exotic eastern elements were indeed rare in the Middlegate floras. Compared with well-dated floras in the Columbia Plateau region, they have the composition of much younger floras of that area, such as the Upper Ellensburg, which is dated at about 10-11 m. y. (C. J. Smiley, 1963). By the same criterion, the Eastgate would also be younger than the late Miocene floras in California to the west, for it has fewer members of the eastern elements than do either the Table Mountain (11 m.y.) or Remington Hill (8 m.y.) floras from the west slope of the Sierra (Condit, 1944a, 1944b). The Eastgate and Middlegate floras also have fewer exotic taxa than the Neroly flora of late Miocene age (10 m.y.) in the Mt. Diablo region (Condit, 1938).

It is apparent that age cannot simply be determined by reference to an age-curve based on a presumed steady rate of environmental change over the western United States during the Tertiary. Forests were changing at different rates in different regions. We need only note that the Creede flora (27 m.y.) of Colorado has 37 genera, 35 of which still live within 160 km (100 mi.) of the fossil locality! By contrast, fully 75% of the genera in the Oligocene Bridge Creek flora (31 m.y.) of eastern Oregon are now exotic to that region, and in the much younger Spokane ("Latah") flora (15 m.y.), 65% of its genera are now exotic to that area. Furthermore, if there were numerous major climatic reversals, as some assert, then the value of any age-curve showing a progressive decrease in the percentage of exotic genera is further vitiated.

Age analysis must be based on sequences of floral change established in <u>local provinces</u> (Axelrod, 1957b). During the Miocene, several vegetation zones characterized the far west, changing with latitude, elevation, and position with respect to the coastal strip (see Axelrod, 1956, fig. 17; 1968, fig. 7). Clearly, local provincial conditions must be weighted carefully if we are to determine age on the basis of plant evidence alone. If we compare the Eastgate with the Blue Mountains flora of eastern Oregon, which is also a conifer-hardwood forest (Chaney, 1959, p. 64), we will naturally assume that the latter is older, inasmuch as it has numerous members of the eastern elements whose derivatives are now exotic to the region. Such a comparison overlooks the critical factor of latitude and its effect on climate; the drier, warmer conditions to the south were not favorable for numerous mesophytic members of the eastern elements. Similarly, if we compare the Eastgate with the Grand Coulee or Spokane floras, which are also somewhat younger (15 m.y.), the differences will lead again to the conclusion that the Eastgate is much younger. However, the floras to the north lived at lower elevation and in a moister climate where a mixed deciduous-broadleaved evergreen forest dominated the lowlands close to sea level. Clearly, in terms of age analysis, comparison of these Nevada floras with those in eastern Washington is analogous to multiplying bananas by oranges.

Similar relations are apparent from a comparison with the 49-Camp flora of northwestern Nevada (LaMotte, 1936; Chaney, 1959, p. 116). Although it has been considered early Miocene (Evernden and James, 1964, p. 1971; Wolfe, 1969, p. 88), this age assignment, like several others (see Axelrod, 1965), was based on an unacceptable radiometric date (Bonham, 1969, p. 19). In addition, geologic mapping shows that the flora has essentially the same stratigraphic position as the nearby Big Basin mammal fauna of early Barstovian age that occurs in beds dated radiometrically at 15-16 m.y. (in Bonham, 1969, p. 18-19). Furthermore, no taxa in the 49-Camp flora are now known to be diagnostic of the early Miocene. The 49-Camp flora has numerous genera that are now found in distant areas, including _Ginkgo_, _Ailanthus_, _Carya_, _Cedrela_, _Cocculus_, _Cercidiphyllum_, _Fagus_, _Nyssa_, _Persea_, _Oreopanax_, _Ulmus_, and _Zelkova_, and they also occur in floras of the nearby region that are dated at 15 m.y. (early Barstovian). Their abundance at 49-Camp is due primarily to its lower elevation and moister climate which supported a rich deciduous hardwood forest near the lakeshore.

In the same way, the differences between the Eastgate and Middlegate floras and the Fingerrock flora (K/Ar age = 16 m.y.), situated 130 km (80 mi.) south (Wolfe, 1964) can be attributed to two factors. First, the Fingerrock flora lived at a lower, warmer (_W_ 14°C) level where there occur exotic hardwood genera (_Carya_, _Cladastris_, _Diospyros_, _Eugenia_, _Ulmus_, _Zelkova_) and species of _Acer_, _Alnus_, _Betula_, _Platanus_, _Mahonia_, and _Populus_ whose nearest relatives are only in areas of ample summer rainfall. The difference in elevation, as judged from the inferred thermal relation, was on the order of 183 m (500 ft.). Second, the Middlegate basin, situated in the lee of higher hills immediately to the west, had less effective summer rainfall. Thus elevation, local terrain, and local climate seem to account for the younger aspect of the Eastgate and Middlegate floras.

A radiometric age of 18.5 ± m.y. for hornblende rhyolite tuff in the middle of the Middlegate Formation indicates a middle Miocene (Hemingfordian) age for the Middlegate and Eastgate floras. This agrees with the Barstovian age (15 m.y.) of a large mammalian fauna which has now been recovered from the basal beds of the Monarch Mill Formation that unconformably overlies the Middlegate Formation about 1 km south of the Eastgate flora. Professor Donald E. Savage has provided the following provisional list of taxa in this fauna, which is now under study:

PRELIMINARY LIST OF MAMMALIAN TAXA IN THE BASAL PART OF THE MONARCH MILL FORMATION, SOUTH OF EASTGATE, NEVADA

	COMMON NAME
INSECTIVORA	
Soricidae	
Trimylus	Shrews
Plesiosoricidae	
Meterix	Shrews
Talpidae	
Mystipterus	Moles
LAGOMORPHA	
Leporidae	
Hypolagus	Rabbits
RODENTIA	
Sciuridae (1 genus)	Squirrels
Castoridae	
Monsaulax	Beavers
Aplodontidae	
Liodontia	Mountain beavers
Mylagaulidae	
Mylagaulus	
Cricetidae, cf. *Peromyscus* .... (2 species)	Woodrats
Heteromyidae (2 genera)	(Mice)
CARNIVORA	
Canidae	
cf. *Tomarctus*	Dogs
Mustelidae (2 genera)	(Weasels, martins)

- cont. -

PERISSODACTYLA	
Equidae	
cf. Hypohippus	Horses
Rhinocerotidae	
Teleoceras	Rhinocerotids
cf. Aphelops	Rhinocerotids
Chalicotheriidae (extinct family)	
ARTIODACTYLA	
Merycoidodontidae	Oreodont
Camelidae	Camels, llamas
Paleomerycidae, cf. Aletomeryx	
Antilocapridae	Pronghorns

It is apparent that these mammals would be at home in the environments that have been reconstructed for the Middlegate basin. The large grazers probably were more frequent in the Middlegate area, where more open country is inferred to have been present (Axelrod, 1956, fig. 17). Near the shore in the Eastgate area, conifer-hardwood forest and sclerophyll forest may have provided shade and rest for the larger grazing mammals, as well as concealment. The rodents and insectivores no doubt occupied several environments, especially riparian and lakeshore, as well as moister parts of the sclerophyll and conifer-hardwood forests. The carnivores ranged widely over the region, searching for prey in all vegetation zones.

REGIONAL RELATIONS

During Miocene time, central Nevada occupied a position transitional between the rich mesic forests of Arcto-Tertiary alliance in the north and the semi-arid woodlands and shrublands of Madro-Tertiary origin to the south (Axelrod, 1940; 1956, fig. 17). In the north, forest composition varied chiefly with elevation (Axelrod, 1965, fig. 12; 1968, fig. 7). Forests in lowland areas (Astoria, Grand Coulee, Lower Ellensburg, Spokane, Clarkia floras) were a rich mixture of evergreen dicots, deciduous hardwoods, and lowland conifers. With a rise in elevation, evergreen dicots disappeared gradually as montane conifers (Abies, Picea, Pinus, Pseudotsuga, Sequoia) assumed a progressively more prominent role in the dominantly deciduous forests (Succor Creek, 49-Camp, Mascall floras) that were gradually transformed into conifer-hardwood forests (Thorn Creek, Blue Mountains, Trapper Creek floras). These finally merged into pure montane conifer forests wherever elevations were near 1,370 m (4,500 ft.) and warmth was about W 11.7°C (53°F) (near Trapper Creek, Thunder Mountain floras). South of central Nevada, there was a rapid decrease in elevation near the present Mohave-Great Basin boundary (lat. 36-37°), falling off into a warm, frostless region dominated by a rich live oak-laurel-palm woodland, with thorn scrub in the warmest sites and pre-chaparral vegetation forming restricted seral stands on drier sites (see Axelrod, 1979, fig. 13). The fossil floras of the northern and southern provinces regularly contain numerous taxa (from 50 to 60%) whose modern allies now are confined to regions with ample summer rainfall.

From a climatic standpoint, the northern area changed from mild temperate to cool temperate climate in upland areas. The region was under the influence of both convectional summer rain and winter cyclonic storms. In the subtropical region to the south, regular convectional showers were present in summer, and hurricanes ("chubascos"), originating in the warmer ocean that then bathed the shores of southern and central California, were no doubt frequent. In the south, winters were comparatively dry, for winter cyclonic disturbances apparently did not then have sufficient strength to penetrate as far south as at present, when high latitudes support a polar (glacial) climate. The position of the Nevada floras in a transitional area between temperate and tropical climates seemingly accounts in part for the unequal distribution of summer rainfall over central Nevada and for the poor representation in many floras of taxa that require ample summer rain.

The west margin of the area certainly was well watered, as shown by the Carson Pass (23 m.y.) and the Sutro (20 m.y.) floras which have numerous summer-rain indicators and indicate a low Sierran divide. Also, floras on the coastward slope of the low Sierra Nevada during the Miocene (Mohawk, Gold Lake, Webber Lake, Niagara Creek) are considerably more mesic than the Nevada floras. By contrast, younger Nevada floras have only a few exotic taxa and these are not well represented quantitatively. Thus the drier summer climate seems consistent with their position in the lee of the low Sierran ridge and low ranges to the west. These older Nevada floras from Middlegate basin do not differ substantially from much younger ones (earlier captioned Mio-Pliocene) such as the Fallon, Chloropagus, Purple Mountain, and Aldrich Station, all now dated radiometrically in the range of 12-13 m.y.

While the low representation of exotic taxa seems partly understandable on the basis of the intermediate position of the region between humid temperate and subhumid subtropical climates to the north and south, and low hills to the west and also along the Sierran axis, occasional floras in the region nonetheless do have a relatively large number of exotic taxa. The Pyramid flora, dated at 15 m.y. and situated 160 km (100 mi.) west, has exotic species of *Acer*, *Aesculus*, *Alnus*, *Betula*, *Crataegus*, *Glyptostrobus*, *Gymnocladus*, *Platanus*, *Pterocarya*, and *Tilia*, and some of them are abundant. Also, the *Lower* Fingerrock flora, situated 120 km (75 mi.) south and dated at 16 m.y., has species of *Acer*, *Betula*, *Carya*, *Diospyros*, *Mahonia*, *Quercus*, *Ulmus*, and *Zelkova* whose nearest allies are in areas with ample summer rainfall, and some of these are subdominants.

The evidence provided by the Eastgate and Middlegate floras suggests that floras with more numerous exotic taxa probably received more abundant and more regular summer rainfall, owing to their favorable topographic position. The Pyramid flora is situated east of the low Sierra and opposite an area where there appears to have been relatively low relief, as judged from geologic mapping in the region (Durrell, 1966). The Fingerrock flora lies east of the Sonora Pass-Ebbetts Pass area, a region that certainly had a relatively low elevation in the middle Miocene. And the

Niagara Creek flora occurs in a 15 m (50 ft.) sedimentary section that rests on the granitic basement below the main accumulation of the Relief Peak Formation and later andesitic units that total 1,500 m in thickness. This flora is dominated solely by deciduous hardwoods, including species of _Alnus_, _Betula_, _Carya_, _Liquidambar_, _Platanus_, and _Populus_. Montane conifers were not recorded, and a single twig of _Taxodium_ implies a relatively lower elevation for the area, as does the composition of the flora itself, which now occurs near the lower margin of fir forest at an elevation of 2,130 m (7,000 ft).

The Ebbetts Pass flora, from a site well up in the subalpine conifer forest at an elevation of 2,490 m (8,200 ft.), is wholly dominated by evergreen dicots, including _Persea_, _Machilis_, and _Magnolia_. It occurs in the basal part of the andesite section, dated at 17 m.y., and about 9 m (30 ft.) above the granitic basement. To the north, the large Carson Pass flora is from andesitic sediments about 20 m above the granitic basement and below andesites dated at 23 m.y. The flora has numerous evergreen dicots that indicate mild winter climate, notably _Actinodaphne_, _Persea_, _Nectandra_, _Ilex_, and _Magnolia_, as well as evergreen oak (cf. _Quercus virginiana_). Among the associated deciduous hardwoods are species of _Crataegus_, _Cyclocarya_, _Fagus_, _Liquidambar_, _Nyssa_, _Platanus_, _Populus_, _Quercus_, and _Ulmus_. No conifers have been found in this deposit which is now at timberline at an altitude of 2,765 m (9,100 ft.).

The evidence suggests that the Pyramid and Lower Fingerrock floras in western Nevada, which have relatively numerous exotic taxa, may have been situated opposite broad, seaward-draining valleys that permitted summer storms to penetrate into the interior from the coastal strip (Axelrod, 1968). The relatively younger aspect of the Middlegate and Eastgate floras, with few exotics, probably reflects a position topographically shielded from these storm corridors. Likewise, local differences in climate over the region, as implied by the younger Miocene floras (Aldrich Station, Horsethief

Canyon, Fallon, Purple Mountain, Chloropagus, Stewart Spring), seem understandable on the basis of terrains that provided slightly different exposures enabling forests or woodlands to contribute larger numbers of taxa to the accumulating record. The more impoverished nature of these younger floras may be related partly to the accumulation of andesites along the Sierran axis, which was now well underway. Although the increase in elevation in the crestal area was only slight, owing to isostatic compensation (Axelrod, 1957a), it probably was sufficient to increase the effectiveness of the axis as a barrier to precipitation from the west and to result also in a greater range of mean monthly temperature, thus giving the region a somewhat less equable climate.

FLORISTIC CHANGES

As is implied in the preceding discussion on the regional distribution of vegetation in west-central Nevada, there is a marked change in the composition of somewhat younger floras in this region. While they still represent the same general vegetation units, notably sclerophyll and conifer-hardwood forest, they are highly impoverished. It thus seems appropriate to review the floras briefly in order to assess the probable factors that may account for the rapid floristic changes that occurred during the period 13-12 m.y. ago.

Middle Miocene Floras

Pyramid Flora

This flora comes from a site 8 km (5 mi.) southwest of Sutcliffe, which is on the west shore of Pyramid Lake and 160 km (100 mi.) northwest of the Middlegate basin. The plants are preserved in diatomite that underlies a basalt which has been dated at 15 m.y. (Bonham, 1969). This flora is dominated by a lake-border deciduous woodland of alder and birch. There are only a few conifers in the flora, indicating that conifer-hardwood forest covered the nearby slopes to the north, where there were sites available on welded rhyolite and quartz latite tuffs that make up the early Miocene (21 m.y.) Hartford Hill Formation. The following are also represented in the Middlegate basin floras by similar or allied (°) species:

Abies laticarpus	Platanus dissecta
Pinus sturgisii	Cercocarpus ovatifolius
Picea sonomensis	Prunus moragensis
Pseudotsuga sonomensis	Sorbus idahoensis
Tsuga mertensioides	Gymnocladus dayana
Populus eotremuloides	Cedrela trainii
Salix pelviga	Acer negundoides
Chrysolepis sonomensis	°Acer osmonti
Lithocarpus nevadensis	Acer oregonianum
Quercus simulata	Acer scottiae
	°Arbutus trainii

°Allied species with larger leaves, or samaras in *Acer*.

The species of *Arbutus*, *Chrysolepis*, *Lithocarpus*, and *Quercus* no doubt formed a sclerophyll forest north of the lake on nearby slopes which faced generally south. The conifer-hardwood forest covered cooler, moister slopes and was filled with hardwoods, including the following that are not represented in the floras from the Middlegate region: *Glyptostrobus oregonensis*, *Pterocarya mixta*, *Ulmus speciosa*, *Acer osmonti*, and *Aesculus* n. sp. Examination of the collection reveals certain features that further emphasize the mesic aspect of this flora. The leaves of the maple, sycamore, alder, and birch are typically large. It is especially noteworthy that the leaves and samaras of *Acer osmonti* are much larger than those of the allied *A. middlegatensis*, consistent with the drier conditions in the interior. The more mesic aspect of this flora seems attributable to a position opposite a seaward-draining, broad valley up which summer storms penetrated from the coastal strip to provide adequate moisture for the deciduous hardwoods that dominate the assemblage.

Lower Fingerrock Flora

This flora comes from a site about a mile north of Stewart Spring, at the northwest corner of the Cedar Mountains, 29 km (18 mi.) northeast of Mina, Nevada, and 120 km (75 mi.) south of the Middlegate basin. The stratigraphic position of this flora is incorrectly shown by Wolfe (1964, fig. 2). It does not underlie a section of lava flows, but comes from shales that rest on dark andesite flows (dated at 16 m.y.) and are overlain by an andesite mudflow breccia also dated at 16 m.y. Above the basal opaline shale is a thin section of fissile shale, the entire sedimentary section totaling scarcely 25 m (80 ft.) thick. The overlying sedimentary section that rests unconformably on the andesite breccia comprises the "Cedar Valley beds" and yields the Stewart Spring flora at sites along the drainage of Fingerrock Wash to the north and west.

The plants at the Fingerrock locality represent two horizons that were grouped into one flora by Wolfe (1964), though they differ considerably in composition. The Upper flora lacks a number of the subdominants of the Lower flora, notably _Quercus pseudolyrata_, _Platanus bendirei_, _Garrya axelrodi_, _Populus lindgreni_, _Acer osmonti_, and others, and includes taxa not present in the Lower flora, notably _Sequoiadendron_. As presently known, the Lower Fingerrock flora has the following taxa in common with the Middlegate and Eastgate floras:

Abies concoloroides	Mahonia reticulata
Picea lahontense	°Amelanchier alvordensis
Picea sonomensis	Cercocarpus ovatifolius
Pinus balfouroides	Heteromeles sonomensis
Pinus sturgisii	Sorbus idahoensis
Chamaecyparis cordillerae	Acer negundoides
Populus cedrusensis	Acer oregonianum
Salix pelviga	°Acer osmonti
Betula thor	Diospyros oregonianum
Quercus hannibali	°Arbutus trainii
Quercus simulata	Rhamnus precalifornica
Mahonia macginitiei	Eugenia nevadensis

°Allied species.

The remains of conifers are not abundant in this assemblage, being represented chiefly by winged seeds that total scarcely 3% of the collection. This implies that a conifer-hardwood forest lived on bordering slopes away from the dominant deciduous hardwoods that covered the lakeshore and nearby slopes. These include the following species that are not recorded in the Middlegate and Eastgate floras:

*Populus lindgreni	Crataegus sp.
Carya bendirei	Sophora spokanensis
Quercus columbiana	(leaflets of 4 unidentified legumes)
*Quercus pseudolyrata	*Acer osmonti
*Ulmus pseudoamericana	Cornus sp.
*Ulmus paucidentata	Garrya axelrodi
*Platanus bendirei	*Zelkova brownii

*Abundantly represented.

There were patches of sclerophyll forest in the area, judging from the abundance of Quercus hannibali, Q. simulata, and Arbutus trainii in the flora. Their associates probably included Cercocarpus ovatifolius, Eugenia nevadensis, Garrya axelrodi, Heteromeles sonomensis, and Rhamnus precalifornica. The community presumably covered warmer, south-facing slopes, bordering the conifer-hardwood forest and the dominant deciduous woodland of the lowlands.

The abundant representation of deciduous hardwoods, including Acer osmonti, Platanus bendirei, Populus lindgreni, Quercus pseudolyrata, Ulmus pseudoamericana, U. paucidentata, and Zelkova brownii, and their absence from the Eastgate-Middlegate area is notable. This is further emphasized by the additional occurrence at Fingerrock of Carya bendirei, Quercus columbiana, and Sophora spokanensis, all typical of floras of the Columbia Plateau region to the north. As is suggested for the Pyramid flora, this locale may have been favorably situated with respect to seaward-draining valleys across the low Sierran divide, so that adequate summer storms penetrated into the area.

Buffalo Canyon Flora

This flora is situated 13 km (8 mi.) southeast of the Eastgate locality and dated at 19 m.y. Many of its species occur in the Middlegate basin floras:

Abies concoloroides
Abies laticarpus
Larix cassiana
Larix nevadensis
Picea lahontense
Picea sonomensis
Tsuga mertensioides
Chamaecyparis cordillerae
Juniperus nevadensis
Populus cedrusensis
Populus eotremuloides
Populus pliotremuloides
Salix desatoyana
Salix pelviga
Salix storeyana
Quercus hannibali
Quercus simulata
Mahonia macginitiei
Mahonia reticulata
Amelanchier grayi
Crataegus middlegatensis
Lyonothamnus parvifolius
Prunus chaneyi
Prunus moragensis
Sorbus idahoensis
Acer negundoides
Acer oregonianum
Acer tyrrelli
Fraxinus coulteri
Arbutus prexalapensis
Eugenia nevadensis

In spite of these similarities, the Buffalo Canyon flora differs from those in the Middlegate basin in two major respects: the remains of conifers are much more abundant at Buffalo Canyon, implying a cooler climate and a somewhat higher elevation, and exotic hardwoods are more abundant in the Buffalo Canyon flora. These include species of Alnus, *Betula, *Carya, Cedrela, Eugenia, Hydrangea, *Juglans, *Prunus, *Ulmus, and *Zelkova, six of which (*) are among its dominant or common species. The differences are understandable, since this flora occupied the windward front of the ancestral Desatoya Range, where greater precipitation would be expected than in the lower Middlegate basin to the west. In addition, the orographic effect of the mountains at the east would tend to give the area higher summer rainfall. Furthermore, the somewhat greater age of the Buffalo Canyon flora also implies that more exotics might be expected there.

Late Miocene Floras

The floras that are younger than 14 m.y. have a markedly different aspect from those of greater age. They are less diverse in taxa, they have fewer species that are summer-rain indicators, the dicots often have smaller leaves than their nearest allies (or ancestors) in slightly older floras, and the floras all have a xeric stamp that implies a shift from humid to subhumid climate. These changes are indicated by a brief review of the younger floras in the region.

Aldrich Station Flora

This flora from the upper middle drainage of the East Walker River, situated on the west front of the Wassuk Range, represents a Sequoiadendron forest that shows affinity with the modern community in its southern areas of occurrence. The following species are also in the Middlegate and Eastgate assemblages or have allied taxa (°) there:

Abies laticarpus	Populus payettensis
Abies concoloroides	Salix storeyana
Picea lahontense	Quercus hannibali
Picea sonomensis	Quercus simulata
Pinus balfouroides	°Betula smithiana
Pinus sturgisii	Betula thor
Pseudotsuga sonomensis	Mahonia macginitiei
Sequoiadendron chaneyi	Platanus paucidentata
Chamaecyparis cordillerae	Cercocarpus antiquus
Populus bonhamii	Lyonothamnus parvifolius
Populus cedrusensis	Rhamnus precalifornica

Absent from the Eastgate and Middlegate floras are the following taxa, of which six are summer-rain indicators (*):

Tsuga sonomensis	*Sophora spokanensis
*Populus subwashoensis	Aesculus ashleyi
*Ulmus moorei	Paxistima myrsinites
*Zelkova nevadensis	*Fraxinus alcorni
Amelanchier apiculata	*Bumelia beaverana
Amorpha oblongifolia	Symphoricarpos wassukana

Chloropagus Flora

This small flora is from the northeast portion of the Hot Springs Mountains that border the Carson Sink, 105 km (65 mi.) northeast of the Middlegate basin. It has the following species in common with the Middlegate and Eastgate floras:

Abies laticarpus	Lithocarpus nevadensis
Picea sonomensis	Quercus hannibali
Pinus balfouroides	Quercus simulata
Chamaecyparis cordillerae	Quercus shrevoides
Juniperus nevadensis	Mahonia reticulata
Populus payettensis	Mahonia trainii
Salix storeyana	Robinia californica
Salix wildcatensis	

Chloropagus taxa absent from the floras near Eastgate are:

Torreya nancyana	Amelanchier apiculata
Populus alexanderi	Cercocarpus linearifolius
Populus subwashoenisis	Cercis carsoniana
Salix payettensis	Ceanothus chaneyi

Fallon Flora

The Fallon flora from the west margin of the Carson Sink, 96 km (60 mi.) west of the Middlegate area, is richly represented by leaves of oak, but relatively poor in species. The following taxa recorded there are also in the Middlegate-Eastgate floras:

Abies laticarpus	Betula thor
Picea sonomensis	Lithocarpus nevadensis
Pinus sturgisii	Quercus hannibali
Chamaecyparis cordillerae	Quercus simulata
Juniperus nevadensis	Quercus shrevoides
Sequoiadendron chaneyi	Mahonia macginitiei
Populus bonhamii	Mahonia trainii
Salix storeyana	

Absent from the Middlegate and Eastgate area are the following:

Torreya nancyana	Sophora spokanensis
Populus subwashoensis	Arbutus matthesii
Salix payettensis	Fraxinus alcorni
Cercocarpus linearifolius	

The summer-rain indicators in this flora include _Betula thor_, _Fraxinus alcorni_, _Mahonia trainii_, _Populus subwashoensis_, and _Sophora spokanensis_. It is amply clear that this assemblage also represents an impoverished conifer-hardwood forest at the upper margin of a sclerophyll forest, the latter dominated by _Quercus_. Drier sites nearby supported a juniper woodland in which were species of _Cercocarpus_, _Mahonia_, and some of the riparian taxa.

As compared with the Middlegate and Eastgate floras, or the more mesic Pyramid and Fingerrock floras, these three late Miocene floras from west-central Nevada (Aldrich Station, Chloropagus, and Fallon) are not rich in taxa. Conifer-hardwood forest was then highly impoverished and restricted over the lowlands at the expense of sclerophyll forest and juniper woodland. Furthermore, there are associated taxa whose nearest relatives are now far to the south or in the interior, implying a warmer and drier climate. Among these are the species of _Aesculus_, _Amelanchier_, _Bumelia_, _Cercis_, _Cercocarpus_, _Fraxinus_, _Juniperus_, and _Mahonia_. Each flora also has only one or two genera, and two or three species, whose living relatives are confined to areas with summer rainfall. The inference that these late Miocene floras imply restricted conifer-hardwood forest over the lowlands, and that richer forests thrived under moister climates in the bordering hills, is supported by the nature of the Chalk Hills flora (Axelrod, 1962), dated at 12 m.y.

Chalk Hills Flora

This flora comes from a site 11 km (7 mi.) northeast of Virginia City and 50 km (32 mi.) west of the Fallon flora. It is preserved in diatomite of the Coal Valley Formation which interfingers with the Kate Peak Formation. The Kate Peak overlies the Chloropagus Formation, in which the Chloropagus flora occurs, and the latter also grades up into the Desert Peak Formation, the basal part of which includes the Fallon flora (for stratigraphic relations see Bonham, 1969, figs. 11 and 12).

The Chalk Hills flora contains 18 species that are in the Middlegate-Eastgate floras or have close allies (°) there:

Abies concoloroides	Salix owyheeana
Pinus sturgisii	°Betula smithiana
Pinus balfouroides	Chrysolepis sonomensis
Pseudotsuga sonomensis	Lithocarpus nevadensis
Sequoiadendron chaneyi	Mahonia macginitiei
Chamaecyparis cordillerae	Ribes stanfordianum
Populus bonhamii	°Amelanchier alvordensis
Populus pliotremuloides	Prunus moragensis
Salix storeyana	Rhamnus precalifornica

Absent from the Middlegate and Eastgate floras are:

Populus washoensis	Ceanothus chaneyi
Salix laevigatoides	Ceanothus leitchii
Carya bendirei	Arbutus matthesii
Holodiscus idahoensis	Rhododendron gianellana

°Allied species.

Two of these, Carya and Populus washoensis, are typical members of Miocene floras to the north. The others all have mesic requirements, as judged from their nearest modern allies. The taxa in common with the Middlegate and Eastgate floras present a more mesic aspect than do those floras. Poplars and birch are especially abundant, and their leaves are large. Note also that the oaks of the sclerophyll forest, *Quercus hannibali* and *Q. shrevoides*, are absent, and that the shrubs are typically those of mesic sites today, notably *Amelanchier*, *Ceanothus*, *Mahonia*, *Ribes*, and *Rhododendron*.

Stewart Spring Flora

The Stewart Spring flora (Wolfe, 1964) comes from sites stratigraphically about 30-80 m above the mammalian fauna of the same name, which is late Barstovian in age (D. Savage, personal communication, 1982). The following species are common to the Middlegate-Eastgate floras:

Abies laticarpus	°Betula smithiana
Picea lahontense	Quercus hannibali
Picea sonomensis	Quercus shrevoides
Pinus sturgisii	Mahonia reticulata
Tsuga mertensioides	°Amelanchier sp.
Chamaecyparis cordillerae	Cercocarpus antiquus
Juniperus nevadensis	°Lyonothamnus cedrusensis
Populus cedrusensis	°Arbutus trainii
Salix pelviga	

°Allied species.

An impoverished conifer-hardwood forest dominated in the bordering hills and valleys, and the lowlands supported sclerophyll vegetation. A number of species give the flora a definitely warmer stamp than the floras of the Carson Sink region 130-160 km (80-100 mi.) northwest. Among these are *Garrya*, *Juglans*, *Peraphyllum*, *Ribes*, *Rhus*, and *Sapindus*. In addition, Wolfe reports two exotic genera, *Astronium* and *Schinus*, which may represent other taxa, and *Colubrina* and *Philadelphus*, which are no more than leaves of *Populus cedrusensis*.

Most of the species representing sclerophyll vegetation in this flora are members of the Madro-Tertiary Geoflora (Axelrod, 1958a; 1979) that wholly dominated the area 320 km

(200 mi.) to the south, as shown by the Tehachapi (Axelrod, 1939), Ricardo (Webber, 1930), Mint Canyon (in Axelrod, 1979), and Anaverde floras (Axelrod, 1950b). It is emphasized that forest conifers (fir, spruce, pine) and their usually associated hardwoods (birch, elm, hickory, broad-leaved oaks, etc.) are entirely absent from these southern Madrean floras that lived under a warmer and drier climate and at a lower altitude. The better representation of sclerophyll taxa at Stewart Spring, as compared with the late Miocene floras to the north, may in part be due to its somewhat lower elevation, and also to a position in the lee of an older terrain in the area directly west, that was elevated following the accumulation of the Fingerrock flora. In addition, the ancestral Sierra to the west probably had a somewhat higher crest than in areas farther north. All these factors would give the Stewart Spring locale a warmer, drier climate, one more suited to sclerophylls than in the area to the north, where the floras of the Carson Sink region indicate more mesic, slightly cooler conditions.

Columbia Plateau Region

A generally similar change in the composition of floras of this region and border areas followed about 14 m.y. ago (Table 13). Inasmuch as these have been reviewed earlier (Chaney, 1959; Axelrod, 1964; Smiley and Rember, 1981), only a brief summary is needed here. The floras of middle Miocene age fall into two different groups in terms of the vegetation and altitude zones that they represent. Relatively lowland floras were dominated by numerous exotic deciduous hardwoods, lowland conifers (_Taxodium_, _Metasequoia_, _Glyptostrobus_), and evergreen dicots of warm temperate requirements (_Cedrela_, _Exbucklandia_, _Machilis_, _Magnolia_, _Persea_). Examples are the Spokane, Clarkia, Grand Coulee, Mascall, and Succor Creek floras. They were succeeded by a series of younger floras, all dated in the range of 13-11 m.y., that have fewer exotic deciduous hardwoods, more temperate taxa with affinities in western North America, and a general absence of evergreen

TABLE 13

Western Floras Show Major Floristic Changes Between 16-14 and 13-11 m.y. Ago

AGE: 16-14 m.y.	13-11 m.y.
Eastern Washington-N. Idaho Spokane, Grand Coulee, Clarkia, Coeur d'Alene	Ellensburg
Eastern Oregon-S. Idaho Trapper Creek, Blue Mts., Sparta, Mascall, Succor Creek	Hog Creek, Thorn Creek, Idaho City, Stinking Water, Almaden, Beulah
Western Great Basin Pyramid, Fingerrock, 49-Camp	Fallon, Purple Mt., Aldrich Stn., Chloropagus, Stewart Spring
Central California Temblor, Niagara Creek	Table Mountain, Neroly
Northern Sierra Nevada Gold Lake, Webber Lake	Denton Creek

dicots, as shown by the Ellensburg, Stinking Water, and Almaden floras; the Ellensburg, nearer the coast, has a few evergreen relicts.

At moderately higher elevations over the region, montane conifers enter the vegetation in larger numbers, the exotic deciduous hardwoods include more numerous species with cool temperate requirements, and evergreen dicots with warm temperate requirements are absent. Examples are the Trapper Creek, Blue Mountains, and Sparta floras. These were succeeded by floras with fewer exotic deciduous hardwoods that require ample summer rainfall, and more taxa that still live in the western United States. Some of these taxa have smaller leaves or samaras and represent distinct ecotypes, distributed in *Acer*, *Amelanchier*, *Betula*, *Fagus*, and others. Examples of these younger floras are the Hog Creek, Thorn Creek, Idaho City, Beulah, and Stinking Water floras.

The younger floras listed in Table 12 have a markedly different aspect. They are less diverse in taxa, have fewer summer rain indicators, the dicots often have smaller leaves than their nearest allies in slightly older floras, there are more numerous species with modern allies in the far west, and the floras have a more xeric stamp that implies a shift from humid to subhumid climate. What changes in environment might have resulted in these floristic modifications?

Late Miocene Environmental Shift

The floras reviewed above show that 13-11 m.y. ago there was a rapid shift to a more subhumid climate over the western interior. This is apparent not only in the central and southern Great Basin but in the Columbia Plateau region as well (Table 12). The trend is reflected in the sharply reduced numbers of broadleaved dicots that are indicators of summer rainfall and in the more modern aspect of the later Miocene floras, all of which have a more xeric stamp.

This trend to a drier climate and to one with less summer rainfall may be related to chilling of the ocean in response to intense, widespread volcanism at this time (Kennett, 1981; Kennett and Thunell, 1975; Kennett, McBirney, and Thunell, 1977; also Scholl et al., 1976), which apparently resulted in the development of a major ice sheet on East Antarctica (Savin et al., 1975; Vogt, 1979; Kennett, 1980; Keller and Barron, 1983). This chilling is recorded along the Pacific coast of North America in the shift from dextrally to sinistrally coiled Globigerina pachyderma during later Mohnian time (Bandy and Ingle, 1970, p. 142, and figs. 2-4; Ingle, 1977). In addition, the diatom flora from the thick Miocene section at Lompoc, California, when compared with the distribution of modern diatoms in surface sediments in the North Pacific, indicates a sharp temperature drop in the late Mohnian followed by a gradual, fluctuating rise during the late Delmontian (Barron, 1973).

Cooler sea-surface temperature would lead to a reduction in summer rainfall and increased drought, and therefore to fewer plants that require summer rain. It is clear that the late Miocene (13-11 m.y.) chilling corresponds in time to the rapid modernization of all presently known western vegetation zones that have a fossil record. The change is not only well marked in the interior, but is apparent also in California, where it is reflected in the marked reduction in summer-rain indicators. The Miocene Temblor flora (15 m.y.) is far richer in mesic deciduous hardwoods than the Table Mountain flora (11-12 m.y.), situated only 190 km (118 mi.) north in the lower foothill belt of the Sierra Nevada. A comparable change is indicated by floras now in the high northern Sierra Nevada, as exemplified by the differences between the Gold Lake (15 m.y.) and Denton Creek (11 m.y.) floras near Blairsden.

SUMMARY

Two contemporaneous middle Miocene (Hemingfordian, 18.5 m.y.) floras situated only 8 km (5 mi.) apart on opposite shores of a lake show major differences in composition. These reflect the role of terrain in controlling local climate to which taxa responded, thus clarifying a major reason for the diverse contemporaneous floras that lived in this province.

The Middlegate flora of 64 species, distributed among 6,882 specimens, covered the southern slopes of a volcanic terrain situated in the lee of hills that have largely been removed by basin faulting and erosion. The dominant taxa formed a sclerophyll forest of *Arbutus*, *Chrysolepis*, *Lithocarpus*, and *Quercus* (3 spp.) that reached close to the lakeshore. Drier slopes supported restricted brushy patches of *Ceanothus*, *Cercocarpus*, *Heteromeles*, *Lyonothamnus*, and *Robinia*. Lake- and stream-border sites were inhabited by *Acer*, *Populus*, and *Salix*. Taxa that contributed to impoverished conifer-hardwood forest lived on distant slopes, reaching down valleys with cold-air drainage toward the lake. These included *Abies*, *Chamaecyparis*, *Picea*, *Pinus*, *Pseudotsuga*, and

Sequoiadendron, associated with *Acer*, *Alnus*, *Betula*, *Crataegus*, *Mahonia*, *Platanus*, *Populus*, *Prunus*, *Salix*, and *Sorbus*. A few rare taxa in the flora have relatives in areas with summer rain, notably the eastern United States and eastern Asia (*Betula*, *Crataegus*, *Diospyros*, *Hydrangea*, *Mahonia*, *Picea*), as well as in the central to southern Rocky Mountains (*Acer*, *Arbutus*, *Cercocarpus*, *Crataegus*, *Populus*).

The Eastgate flora of 55 species, distributed among more than 5,800 specimens, accumulated near an active fault scarp along the west front of the Eastgate Hills. This steep, hilly area on the windward side of the lake had both north- and south-facing valley walls close to the shore. Current-direction analysis suggests some transport of plant material from the south, where cool north-facing slopes also reached the lake. The abundant taxa contributed to conifer-hardwood forest, with a canopy of *Abies*, *Larix*, *Picea*, *Pinus*, *Pseudotsuga*, *Tsuga*, *Chamaecyparis*, and *Sequoiadendron*, and an understory of *Acer*, *Aesculus*, *Amelanchier*, *Betula*, *Chrysolepis*, *Crataegus*, *Mahonia*, *Prunus*, *Ribes*, and *Sorbus*. Warm, south- and west-facing slopes supported patches of sclerophyll forest dominated by *Arbutus*, *Eugenia*, *Lithocarpus*, and *Quercus*; they were also important in the lower conifer-hardwood forest, much as their descendants are today. Sclerophyll shrubs included *Cercocarpus*, *Chrysolepis*, *Heteromeles*, and *Mahonia*, and together with some deciduous taxa, such as *Amelanchier*, *Crataegus*, and *Prunus*, probably formed local brushy patches of seral nature. The lakeshore supported a deciduous woodland of *Acer*, *Betula*, *Fraxinus*, *Populus*, and *Salix*, as well as the near-hydric conifers *Chamaecyparis* and *Sequoiadendron*, and the mesic *Lithocarpus* and *Quercus* among the sclerophylls. Most of the few taxa whose nearest allies are now in areas with summer rain, notably *Acer*, *Aesculus*, *Arbutus*, *Betula*, *Eugenia*, and *Mahonia*, have a low quantitative representation.

The climate is mild temperate in areas where modern vegetation includes numerous taxa allied to those in these fossil floras, notably the forests on the western slope of the Sierra Nevada and in the Coast Ranges. Taking diverse paleoclimatic factors into account, it is inferred that mean annual temperature (_T_) was near 12°C and the mean monthly range of temperature was about 15°C, giving a warmth of _W_ 13.3°C, or a growing season of 162 days with mean temperature warmer than that. Equability was near _M_ 58, as compared with _M_ 50-55 in most related modern forests and _M_ 47 in the Middlegate basin today. Annual rainfall was near 760-890 mm at Middlegate, as compared with 890-1015 mm at Eastgate. A few rare exotic taxa in each flora imply some summer rainfall, probably not more than 50-75 mm each summer month. Frost frequency was about 4-5% of the hours of the year, so the area received occasional light dustings of snow.

Comparison of thermal conditions inferred for the floras with that estimated for Miocene floras at sea level in central California suggests a difference in _T_ of about 5°C. Assuming a near-normal terrestrial lapse rate (183 m/1°C), the elevation of the lake was near 915 m (3,000 ft.).

The middle Miocene age of these floras is indicated by a radiometric date of 18.5 m.y. on a rhyolitic tuff stratigraphically just below the floras. A mammalian fauna of 20-odd taxa in the basal beds of the Monarch Mill Formation, which uncomformably overlies the Middlegate Formation, is early Barstovian (15 m.y.).

The marked differences between the contemporaneous Eastgate and Middlegate floras reflect their local settings. The Middlegate flora covered south-facing slopes in the lee of hills to the west, whereas the Eastgate was on the windward side of the lake, bordered by steep hills with cooler, north-facing slopes that provided mesic sites near the lake. As a result, sclerophyll forest dominated at Middlegate, conifer-hardwood forest at Eastgate. These floras differ from others in nearby Nevada (i.e., Buffalo Canyon, Pyramid, Lower Fingerrock) that have more numerous exotic taxa. The latter two floras may have been situated opposite low topographic corridors along the Sierran axis to the west, where humid climates supported rich mesophytic forests dominated by exotic taxa. By contrast, geologic evidence shows that the Middlegate and Eastgate floras were in the rainshadow of a moderate range which has since largely disappeared as a result of downfaulting and erosion. Greater differences are apparent with floras farther north in Oregon, Idaho, and Washington, where up to 60% exotic taxa are recorded, and also to the south in the drier Madrean province, where ample summer rainfall and frostless climate prevailed. The regional differences in composition seem to reflect a lower and more uneven distribution of summer rainfall over central Nevada, a feature controlled partly by local terrain as well as by its position between contrasting humid temperate and semi-arid subtropical climates to the north and south, respectively.

There was a marked change in floral composition over the entire region, including the Columbia Plateau, 13-11 m.y. ago. This seems attributable to rapid ocean chilling as indicated by planktonic foraminifera and marine diatom floras. The rapid lowering of ocean temperature reflects the spread of the East Antarctic ice sheet, which may have resulted from intense volcanism (Columbian Episode). Chilling led to a significant decrease in summer rain and hence to a marked reduction in the exotic taxa which largely dominate the older floras. The younger floras also have a more subhumid aspect and are composed of more numerous taxa whose nearest relatives are now in the western United States.

SYSTEMATIC DESCRIPTIONS

(All specimens are deposited in the
Museum of Paleontology,
University of California, Berkeley)

Family EQUISETACEAE

Equisetum alexanderi Brown

Equisetum alexanderi Brown, U.S. Geol. Surv. Prof. Paper 186-J, p. 167, pl. 45, fig. 5, 1937.

Axelrod, Univ. Calif. Pub. Geol. Sci., vol. 33, p. 275, pl. 25, fig. 1, 1956.

Apart from the two stem fragments collected initially at the Middlegate site, no additional material of this small-stemmed horsetail has been recovered there. Equisetum regularly inhabits moist sites, as along sandbars in rivers and in swampy areas at the margins of lakes.

Occurrence: Middlegate, hypotype no. 4252, homeotype no. 4253.

Family PINACEAE

Abies concoloroides Brown

(Plate 16, figs. 1-4)

Abies concoloroides Brown, Wash. Acad. Sci. Jour., vol. 30, p. 347, 1940.

Axelrod, Univ. Calif. Pub. Geol. Sci., vol. 33, p. 274, pl. 12, figs. 6-8, 1956.

Axelrod, Univ. Calif. Pub. Geol. Sci., vol. 39, p. 227, pl. 42, figs. 1-3, 1962.

Axelrod, Univ. Calif. Pub. Geol. Sci., vol. 51, p. 105, pl. 5, figs. 4-8 only; pl. 7, fig. 1, 1964.

The needles of this species are relatively long and the winged seeds are typically wedge-shaped, with the wings attached to the upper part of the seeds. The cone scales are not as convex distally as those of the rather similar *A. laticarpus* and its modern analogue, *A. magnifica* Murray. The winged seeds of *A. laticarpus* are more asymmetrical and the wings are attached farther down on the seed than those of *A. concoloroides*.

The structures of *A. concoloroides* are similar to those produced by the living white fir, *A. concolor* Lindley and Gordon, a widely distributed tree in the western United States. In California (dry summers), white fir is restricted to the middle and upper part of the mixed-conifer forest. However, in the Rocky Mountains and border areas of the Colorado Plateau and eastern Great Basin, where there is summer rainfall, *A. concolor* descends into the lower part of the forest and in favorable areas occurs within a stone's throw of pinyon-juniper woodland vegetation.

Occurrence: Eastgate, hypotype nos. 6684-87, homeotype nos. 6688-89.

Abies laticarpus MacGinitie
(Plate 4, figs. 1-5; pl. 16, figs. 5-12)

Abies laticarpus MacGinitie, Carnegie Inst. Wash. Pub. 416, p. 47, pl. 3, fig. 5, 1933.

Chaney and Axelrod, Carnegie Inst. Wash. Pub. 617, p. 139, pl. 11, fig. 8, 1959.

Abies concoloroides Axelrod, Univ. Calif. Pub. Geol. Sci., vol. 33, p. 275, pl. 4, figs. 2-6; pl. 12, fig. 6; pl. 17, figs. 5, 6; pl. 25, fig. 5, 1956.

Axelrod, Univ. Calif. Pub. Geol. Sci., vol. 51, p. 105, pl. 5, figs. 9, 10 only, 1964.

Abies concolor Wolfe, not Lindley, U.S. Geol. Surv. Prof. Paper 454-N, p. N4, pl. 1, fig. 10; pl. 6, figs. 1-3, 6, 10, 11, 1964.

Abies sp. Wolfe, ibid., p. N14, text-fig. 5, 1964.

As noted above, this species is distinguished from _A._ _concoloroides_ on the basis of its cone scales, which are more concave distally, and by the seed wing, which is more asymmetrical and broader and is attached lower down on the seed, not distally or nearly so.

These remains are similar to those produced by red fir, _A._ _magnifica_ Murray and the var. _shastensis_ Lemmon, handsome trees that occur chiefly in the upper part of the mixed-conifer forest and in the higher red fir and subalpine forests. In the southern Sierra Nevada (Tulare County), _A._ _shastensis_ reaches well down into the mixed-conifer forest and is not far from sclerophyll vegetation. Under the Miocene climate of summer rainfall, _A._ _laticarpus_ evidently lived at lower levels than its relative does today. This inference is consistent with its relatively common occurrence in the Eastgate, Middlegate, and nearby late Miocene floras, where it is associated with remains of _Cercocarpus_, _Lithocarpus_, _Quercus_, and others that are far removed from it today.

Occurrence: Middlegate, hypotype nos. 6400-04, homeotype nos. 6405-10; Eastgate, hypotype nos. 6690-97, homeotype nos. 6698-6704.

Abies _scherri_ Axelrod
(Plate 4, figs. 14-15)

Abies _scherri_ Axelrod, Missouri Bot. Garden Ann., vol. 63, p. 28, figs. 4, 5b, 6, 8, 1976.

Torreya _nancyana_ Axelrod, Univ. Calif. Pub. Geol. Sci., vol. 33, p. 281, pl. 18, fig. 6 only, 1956.

The record of this species in the Middlegate flora consists only of a sharply pointed needle with twisted petiole and a small seed with a short wing attached distally.

The fossils are similar to those produced by the living _A._ _bracteata_ (D. Don) Nuttall (=_A._ _venusta_ K. Koch), which is sufficiently distinct to represent the sole member of the subgenus _Pseudotorreya_, confined now to scattered areas in the Santa Lucia Mountains south of Monterey, California. It is a member of the sclerophyll forest, reaching up into the lower scattered groves of mixed-conifer forest.

Fossils previously compared with *A. bracteata* represent other taxa. *A. longirostris* Knowlton from Creede, Colorado (Knowlton, 1923), is an extinct member of a presently Asian group of species. *A. chaneyi* Mason (Mason, 1927) represents two taxa, the foliage being *Cephalotaxus* and the winged seeds *A. longirostris* (in Axelrod, 1976b).

The above-cited specimen of *Torreya* from the Fallon flora displays the prominently twisted petiole typical of *A. bracteata* and is therefore transferred to *A. scherri*. However, the specimens of *T. nancyana* from the Chloropagus flora (Axelrod, 1956, pl. 12, figs. 13-16), have a petiole like *Torreya* and are thus retained as *T. nancyana*.

Occurrence: Middlegate, paratype nos. 5491-92, homeotype type no. 6411.

Larix cassiana Axelrod
(Plate 17, figs. 1-2)

Larix cassiana Axelrod, Univ. Calif. Pub. Geol. Sci., vol. 51, p. 107, pl. 6, figs. 1-6, 1964.

Three winged seeds in the Eastgate flora are similar to those recovered at Trapper Creek, Idaho. They are much like those produced by *L. potaninii* Batalin, a common tree in the conifer forests of China, from Shensi to the Tibetan borderland. The seeds are relatively large and have a large but short, rounded wing, and thus differ from seeds of most other living species of larch.

It is noteworthy that in both the Trapper Creek and Eastgate floras, *L. cassiana* is associated with numerous montane conifers and broadleaved sclerophyll genera are also present. They presumably occupied drier sites on volcanic slopes bordering the areas of plant deposition.

The Trapper Creek flora occurs in sediments that lie with *angular discordance below* the Idavada Volcanics and Jenny

Creek Tuff, and is therefore older, contrary to Mapel and Hail (1959). No evidence has been presented to support the opinion (Wolfe, 1969; Leopold and MacGinitie, 1972) that it is late Miocene (see discussion in Axelrod, 1964). Nor is the notion acceptable that pollen of Ephedra and Sarcobatus recorded in the sediments indicates that they were part of the flora (Wolfe, 1969). They could not possibly have lived in an area with high rainfall, as is indicated by the dominant montane conifers and deciduous hardwoods. It is more probable that the pollen was transported northward by summer storms from southern Nevada and border areas where there was dry climate at that time (Axelrod, 1979); that pollen of Ephedra and Sarcobatus may be carried long distances from the areas of their natural occurrence is well documented (Maher, 1964).

Occurrence: Eastgate flora, hypotype nos. 6705-06, homeotype no. 6707.

Larix nevadensis Axelrod, new species
(Plate 17, figs. 11-14)

Description: Winged seeds averaging 9-12 mm long; seed relatively large, acutely rounded distally; wing tapering upward distally to an acute or blunt tip. Filamentous needles not found.

Discussion: The winged seeds of this larch are relatively common at the Eastgate site and are also in the nearby Buffalo Canyon flora. They are easily separated from the larger seeds of Tsuga, which have a more nearly rectangular wing that is quite wide under the seed and the seeds have conspicuous resin deposits that appear on the fossil. Picea seeds differ from those of Larix in having an asymmetrical wing widest distally, and the seed tips are snubbed.

L. occidentalis Nuttall produces winged seeds similar to these fossils. It occurs in the northern Rocky Mountains of Idaho-Montana-British Columbia and reaches into the eastern Cascades of Washington and northern Oregon. In much of its area rainfall totals at least 150-200 mm (6-8 in.) in the three summer months.

The unidentified seed (Wolfe, 1964, pl. 6, fig. 25), which I earlier identified as Larix and named Larix cedrusensis (Axelrod, 1981, p. 211), is either an aborted seed of Tsuga or, more probably, is indeterminate; hence the name L. cedrusensis is disposed of.

Occurrence: Eastgate, holotype no. 6708, paratype nos. 6709-11.

Picea lahontense MacGinitie
(Plate 4, figs. 8-10; pl. 17, figs. 7-10)

Picea lahontense MacGinitie, Carnegie Inst. Wash. Pub. 416, p. 46, pl. 3, figs. 6, 8 (not fig. 4, which is Picea sonomensis Axelrod), 1933.

Pseudotsuga masoni MacGinitie, ibid., p. 47, pl. 3, figs. 1-3, 1933.

Picea magna MacGinitie, Carnegie Inst. Wash. Pub. 599, p. 83, pl. 18, figs. 5-7, 1953.

Axelrod, Univ. Calif. Pub. Geol. Sci., vol. 33, p. 275, pl. 4, figs. 7-12; pl. 25, figs. 8, 9, 1956.

Chaney and Axelrod, Carnegie Inst. Wash. Pub. 617, p. 140, pl. 12, figs. 10-15, 1959.

Axelrod, Univ. Calif. Pub. Geol. Sci., vol. 51, p. 108, pl. 6, figs. 9-13, 1964.

Wolfe, U.S. Geol. Surv. Prof. Paper 454-N, p. N15, pl. 1, figs. 3, 5; pl. 6, figs. 7, 12, 17, 18, 22, 1964.

Picea breweriana Wolfe, not S. Watson, ibid., p. N14, pl. 6, figs. 4, 5, 8, 9, 13 (figs. 14 and 19 are P. sonomensis Axelrod), 1964.

Pinus harneyana Chaney and Axelrod. Axelrod, Univ. Calif. Pub. Geol. Sci., vol. 51, p. 108, pl. 6, figs. 20-22 (fig. 19 is Pinus fascicle), 1964.

In discussion with Howard Schorn on the status of winged seeds referred to Pinus and Picea, he pointed out that Picea seeds have parallel veins the length of the wing, whereas those of Pinus develop a characteristic undulating cross-ridging due to compression during the two-year period of maturation.

Large suites of winged seeds show that there is a complete range in shape from the wings of Picea magna, which are broadened distally, to those that are not conspicuously widened and simulate Pinus; the seeds of both have the prominent snubbed nose of the species. We have therefore concluded that all the above-listed material represents a species of spruce that is probably extinct. As noted earlier, the seeds show some relationship to Picea polita, but the needles, of which the Trout Creek leafy twig is the only specimen known to us (MacGinitie, 1933, pl. 3, fig. 8), are at least twice as long as those of P. polita; they are more slender and not so sharply pointed.

Occurrence: Middlegate, hypotype nos. 6412-14, homeotype nos. 6415-20; Eastgate, hypotype nos. 6721-24, homeotype nos. 6725-37.

Picea sonomensis Axelrod
(Plate 4, figs. 6-7; pl. 17, figs. 3-6)

Picea sonomensis Axelrod, Carnegie Inst. Wash. Pub. 553, p. 190, pl. 36, fig. 2; p. 251, pl. 42, figs. 2, 3, 1944.

Axelrod, Univ. Calif. Pub. Geol. Sci., vol. 33, p. 276, pl. 4, figs. 13-16; pl. 12, fig. 5; pl. 17, figs. 7-9; pl. 25, figs. 6, 7, 1956.

Chaney and Axelrod, Carnegie Inst. Wash. Pub. 617, p. 141, pl. 12, figs. 4-9, 16, 17, 1959 (see synonymy).

Axelrod, Univ. Calif. Pub. Geol. Sci., vol. 51, p. 108, pl. 6, figs. 14-18, 1964.

Picea breweriana Wolfe, not S. Watson, U.S. Geol. Surv. Prof. Paper 454-N, p. N14, pl. 6, figs. 14 and 19 only (not figs. 4, 5, 8, 9, 13, which are Picea lahontense Macg.), 1964.

The seeds of *P. sonomensis* are distinguished by a wing that is markedly broadened distally and a small, snubnosed seed. They are thus readily separated from those of *P. lahontense* (=*P. magna*), with which it often occurs.

Among living species, the winged seeds produced by *P. breweriana* S. Watson of northwestern California are quite similar. This spruce occurs in the upper mixed-conifer forest and in the fir-hemlock forest above it. It is amply clear from its fossil associates that *P. sonomensis* had a much wider occurrence, reaching down into the lower part of the mixed-conifer forest during the Miocene and later, a distribution that probably was made possible by the regular incidence of summer rainfall.

Occurrence: Middlegate, hypotype nos. 6421-22, homeotype nos. 6423-31; Eastgate, hypotype nos. 6738-41, homeotype nos. 6742-56.

Pinus alvordensis Axelrod
(Plate 17, figs. 15-19)

Pinus alvordensis Axelrod, Carnegie Inst. Wash. Pub. 553, p. 251, pl. 42, fig. 4, 1944.

In the Eastgate flora there are a number of small ovate seeds with long, slender wings which are similar to the Alvord Creek pine and resemble the winged seeds of the living *P. contorta* Douglas. This tree is found most commonly at higher elevations in the Sierra Nevada and Cascades to the north, but in the Rocky Mountains it descends to lower elevations to enter the mixed-conifer forests composed of *Pseudotsuga*, *Larix*, *Pinus* (*scopulorum*, *flexilis*), *Abies concolor*, and their regular associates. A similar relationship is suggested for the Eastgate flora, where *P. alvordensis* occurs with the allies of these taxa and their associates.

Occurrence: Eastgate, hypotype nos. 6757-61, homeotype nos. 6762-69.

Pinus balfouroides Axelrod
(Plate 16, figs. 13-14)

Pinus balfouroides Axelrod, Univ. Calif. Pub. Geol. Sci., vol. 121, p. 209, 1980.
Pinus florissantii Lesquereux. Axelrod, Univ. Calif. Pub. Geol. Sci., vol. 39, p. 227, pl. 42, fig. 9, 1962.
Pinus wheeleri Cockerell. Axelrod, ibid., p. 227, pl. 42, figs. 4-8, 1962.
Pinus quinifolia Smith. Axelrod, Missouri Bot. Garden Ann., vol. 63, p. 28, figs. 13-14, 1976.

The above-listed fossils from the Chalk Hills and Purple Mountain floras are allied to the living *P. balfouriana* Greville and Balfour, as judged from cones, winged seeds, and fascicles in those floras. Similar winged seeds and fascicles in the Eastgate flora seem inseparable from them and are therefore referred to this species.

P. balfouriana has a discontinuous distribution, occurring in the Klamath Mountain region of northwestern California and in the southern Sierra Nevada, 640 km (400 mi.) southeast. In its northern area it reaches down into the mixed-conifer forest from higher levels, but in the south it is restricted to levels well above 2500 m. The northern and southern populations are quite distinct (Mastrogiuseppe, 1980), and presumably originated since 12 m.y. ago as the former area of *P. balfouroides* in western Nevada (and California) was reduced by spreading drought.

Occurrence: Eastgate, hypotype nos. 6770-71, homeotype nos. 6672-76.

Pinus sturgisii Cockerell
(Plate 4, fig. 17; pl. 17, figs. 25-26)

Pinus sturgisii Cockerell, Amer. Jour. Sci., ser. 4, vol. 26, p. 538, text-fig. 2, 1908.
Pinus florissantii Lesquereux. MacGinitie, Carnegie Inst. Wash. Pub. 599, p. 84, pl. 18, fig. 12; pl. 20, figs.

1, 3, 4 (not pl. 19, fig. 2, which remains _P. florissantii_ Lesquereux), 1953.

Axelrod, Univ. Calif. Pub. Geol. Sci., vol. 33, p. 276, pl. 4, figs. 19, 20; pl. 17, figs. 10, 11, 1956.

Several broken pine needles and a few battered winged seeds in the collections are referred to this species. The material is well matched by the needles and seeds of the living _P. ponderosa_ Lawson, a widely distributed tree in the western United States. Needle length is quite variable, depending on the region in which the trees grow.

As noted earlier (Axelrod, 1980, p. 209), the type specimen of _P. florissantii_ Lesquereux is a cone similar to those of _P. flexilis_ James.

Occurrence: Middlegate, hypotype no. 6432, homeotype nos. 6433-40; Eastgate, hypotype nos. 6777-78, homeotype nos. 6779-87.

Pseudotsuga sonomensis Dorf

(Plate 4, fig. 16; pl. 16, figs. 15-19)

Pseudotsuga sonomensis Dorf, Carnegie Inst. Wash. Pub. 412, p. 72, pl. 6, figs. 2-4, 1930.

Axelrod, Carnegie Inst. Wash. Pub. 553, p. 191, pl. 36, fig. 3, 1944.

Axelrod, ibid., p. 251, pl. 42, fig. 1, 1944.

Chaney and Axelrod, Carnegie Inst. Wash. Pub. 617, p. 143, pl. 13, figs. 13-15, 1959.

Axelrod, Univ. Calif. Pub. Geol. Sci., vol. 39, p. 228, pl. 42, figs. 10-11, 1962.

Axelrod, Univ. Calif. Pub. Geol. Sci., vol. 51, p. 110, pl. 6, figs. 26-30, 1964.

Axelrod, Missouri Bot. Garden Ann., vol. 63, p. 28, figs. 7-9, 1976.

Numerous winged seeds and several needles represent this species in the Eastgate flora, but the Middlegate site has yielded only a few of its seeds. These relations seem consistent with the topographic setting of the floras, with the former in cooler, moister sites.

The winged seeds of Pseudotsuga are relatively large, the wing is markedly acute distally, and the seed is truncated sharply at an acute angle where it adheres to the wing. The slender needles are readily distinguished by their acute to rounded tips and constriction at the base into a distinct petiole.

The fossils are similar in all respects to the foliage, winged seeds, and cones produced by P. menziesii (Mirbel) Franco. This widely distributed tree in the western United States ranges northward into the mountains of Alberta and British Columbia, but does not reach into the Alaska panhandle. The Rocky Mountain form, var. glauca (Beissener) Franco, is not sufficiently distinct to be recognized in the present material.

Occurrence: Middlegate, hypotype no. 6441, homeotype no. 6442; Eastgate, hypotype nos. 6788-91, homeotype nos. 6792-6803.

Tsuga mertensioides Axelrod
(Plate 4, figs. 11-13; pl. 16, figs. 20-23)

Tsuga mertensioides Axelrod, Univ. Calif. Pub. Geol. Sci., vol. 51, p. 110, pl. 7, figs. 4-12, 1964.

Tsuga heterophylla Wolfe, not Sargent, U.S. Geol. Surv. Prof. Paper 454-N, p. N15, pl. 6, figs. 15, 16, 20, 21, 24, 1964.

The winged seeds of this species are rare in the Middlegate flora but relatively common at the Eastgate and Buffalo Canyon sites. However, no needles or cones of hemlock were recovered at any of these localities. This suggests that it was confined chiefly to the middle and upper parts of the conifer-hardwood forest in the bordering hills.

The winged seeds are readily recognized by their relatively large size and by the wing, which is essentially rectangular in outline.

The seeds referred to _T_. _heterophylla_ Sargent by Wolfe are much larger than those produced by that species but similar to normal winged seeds of _T_. _mertensiana_ (Bongard) Carriere. The fossils are not separable from the seeds produced by the living species but are given fossil names for several reasons, as outlined previously (Axelrod, 1980, p. 205-206).

Occurrence: Middlegate, hypotype nos. 6443-45; Eastgate, hypotype nos. 6804-07, homeotype nos. 6808-20.

Family CUPRESSACEAE

(Contributed by Stephen W. Edwards, Paleontology Dept., University of California, Berkeley, from part of a Ph.D. thesis)

Chamaecyparis cordillerae Edwards and Schorn, new species

(Plate 4, fig. 21; pl. 17, figs. 21, 23; pl. 23, figs. 4-5)

Thuja _dimorpha_ auct. non. (Oliver) Chaney and Axelrod. Axelrod, Univ. Calif. Pub. Geol. Sci., vol. 33, p. 278, pl. 12, figs. 1-4; pl. 18, figs. 1, 2; pl. 25, figs. 2, 3, 1956.

Chamaecyparis _linguaefolia_ auct. non. (Lesq.) MacGinitie. Axelrod, Univ. Calif. Calif. Pub. Geol. Sci., vol. 39, p. 229, pl. 43, figs. 1, 2, 6, 1962.

Chamaecyparis _nootkatensis_ (D. Don) Spach. Wolfe, U.S. Geol. Surv. Prof. Paper 454-N, p. N15, pl. 6, figs. 27, 30, 31, 34-37, 1964.

Chamaecyparis _sierrae_ Axelrod. Axelrod, Missouri Bot. Garden Ann., vol. 63, p. 31, figs. 10, 11, 1976.
Axelrod, Univ. Calif. Publ. Geol. Sci., vol. 121, p. 207, 1980.

Many specimens of Cupressaceae have been recovered from the Eastgate and Middlegate deposits. Some are definitely referable to _Chamaecyparis_, and some are _Juniperus_. A number of fragmentary specimens cannot at present be confidently referred to either of these genera, although probable assignments can be made for nearly all specimens on the basis of criteria derived from studying extensive samples (Edwards, 1983). Uncertainty of generic identification of some foliar specimens from these and other localities of the Nevada Miocene results from the overlapping morphologic ranges of the two genera. The fossil *Chamaecyparis* is allied to *C. nootkatensis*, morphologically the most juniper-like living member of the genus; and the Miocene taxon is even more juniper-like than the living one. Moreover, the fossil *Juniperus*, *J. nevadensis*, bears its leaves in decussate pairs (as *Chamaecyparis* does) more frequently than in ternate whorls. Some specimens show decussate phyllotaxis through three orders of shoots. (The smallest, ultimate shoots on a herbarium-sheet-sized spray are collectively called order no. 1; the next-smallest group is order no. 2, and so on.)

The Eastgate and Middlegate *Chamaecyparis* is represented by flat-sprayed foliage, cupressoid cones, and biwinged seeds. It is inseparable from the taxon represented by extensive collections from Stewart Valley, Nevada (UCMP locs. PA 203 and PA 327), and Chalk Hills, Nevada (UCMP loc. P3526). This material is distinguishable from the associated juniper because the latter: (1) bears ternate whorls on some shoots; (2) displays on some shoots twisting of the axis and/or folding over of branches, resulting from pressing of foliage that was not flattened in life; and (3) where ternate whorls or evidence of nonflattened foliage are lacking, specimens referable to *Juniperus* are far more openly branched (skipping more pairs of lateral leaves between branches), and more uniform in leaf morphology from order to order, than *Chamaecyparis*. *Juniperus* lower-order (nos. 1 and 2) shoots also show a greater tendency to pronounced curvature or sharp directional changes.

C. cordillerae differs from *C. nootkatensis* in its seeds: the wings are larger in the Miocene species (more *Callitris*-like); the resin vesicles are multiple (usually lacking but sometimes present and dual in _nootkatensis_); and the basal end of the seed body is more consistently and more extensively truncate. In addition, the foliage of _C. cordillerae_ differs from that of _C. nootkatensis_ in many characters. Chief among these are: (1) in _cordillerae_, the leaf whorls on lower-order (nos. 1 and 2) shoots are usually squatter, and the facials more sharply pointed; (2) the facial-leaf shoulders on higher-order (nos. 3-5) shoots are sharper, and the portion of the facial distal to the shoulders is less elongate; (3) facial non-overlap in higher shoot orders is more consistent and pronounced than in _nootkatensis_; and (4) on higher-order shoots, the conspicuous hollowing of facial leaves proximal to their shoulders where branches arise, standard in _nootkatensis_, is poorly developed.

C. cordillerae differs from _C. lawsoniana_ (A. Murr.) Parl. in its seeds, which have a long-ovate or amygdaloidal (almond-shaped) body (rather than elliptical as in _C. lawsoniana_), broader wings, multiple rather than dual resin vesicles, and a more truncate basal end; and in its foliage, which is more similar to _C. nootkatensis_, with its greater acuteness of facial-leaf tips, more consistent overlap of facial leaves, greater flare of lateral-leaf tips, and greater frequency of skipped lateral-leaf pairs between branches in higher-order shoots. Some fragmentary specimens from Stewart Valley and one from Eastgate are similar in some ways to _C. lawsoniana_, but they are isolated specimens in a large sample which is unique. It is possible to find rather _lawsoniana_-like shoots or sprays on some trees in living _nootkatensis_ groves.

Female cones are not well preserved at the Nevada localities, but the available specimens suggest a low cone-scale number, as in _C. nootkatensis_ (usually 4-6), in contrast to _C. lawsoniana_ (usually 8 or more). Rouane (1973, p. 200) and Owens and Molder (1974, p. 2079) noted that in _C. nootkatensis_ apical vegetative growth is frequently resumed at the same apex where the pollen cone has been shed, but not in _C. lawsoniana_. Such resumption of apical growth seems to be

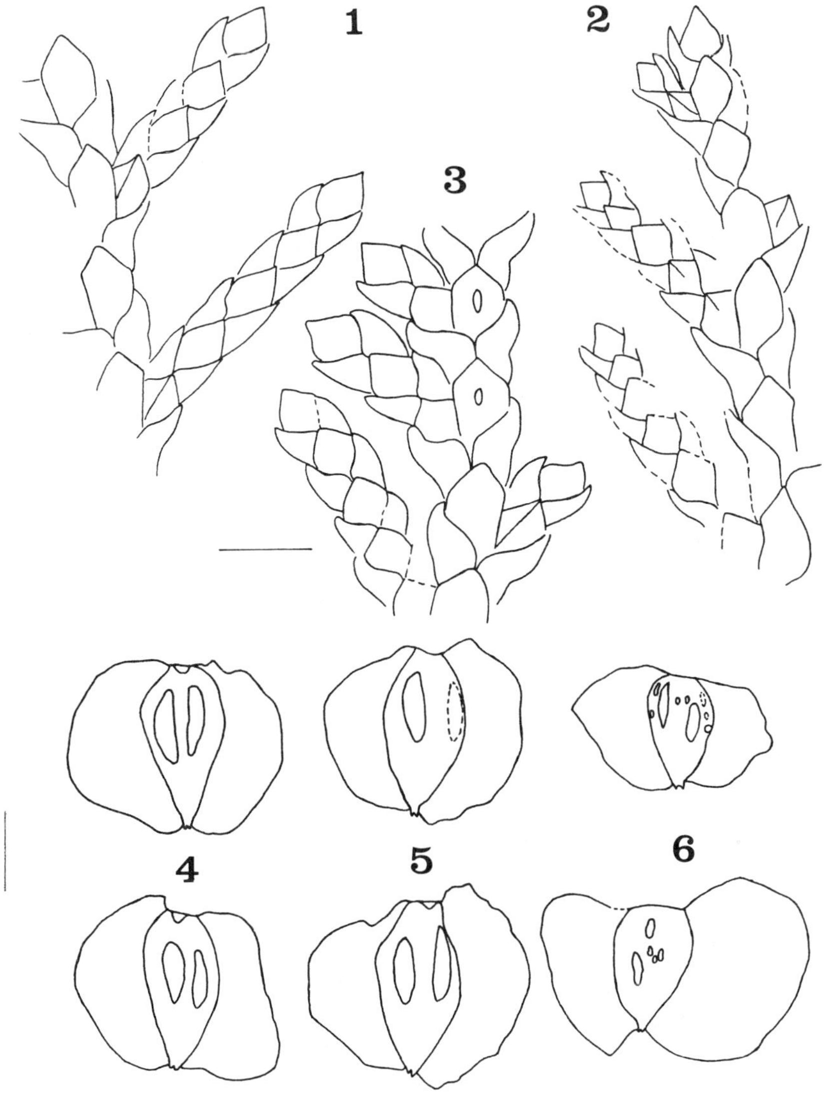

FIGURE 5. Foliage and seeds of *Chamaecyparis cordillerae* Edwards and Schorn, compared with those of the living *C. nootkatensis* and *C. lawsoniana*. Foliage of (1, 2) Middlegate, (3) Eastgate; drawings are partial representations of original specimens. Seeds of (4) *C. nootkatensis* from timberline, Mount Rainier; (5) *C. lawsoniana* from University of California, Berkeley, Botanical Garden; (6) *C. cordillerae* from Eastgate. Drawings by Steven W. Edwards. Bar scale = 2 mm.

present on the largest, most extensive spray from Stewart Valley (UCMP loc. PA 327).

Axelrod (1980, p. 207) noted that the Nevada Miocene *Chamaecyparis* was more like *C. lawsoniana* than *C. nootkatensis* because it bears glands on the facial leaves as in the former species. Glands are indeed common in the fossils, but not as frequent as in *C. lawsoniana*. *C. nootkatensis* does bear glands occasionally. Wolfe (1964, p. N15) referred the Nevada Miocene material to *C. nootkatensis*, but the fossils are better designated as a distinctive, extinct species.

Occurrence: Stewart Spring, holotype USNM 42024; paratype UCMP 8032, Chalk Hills; Middlegate, hypotype no. 6542, homeotype nos. 6253-55; Eastgate, hypotype nos. 6842-44, 6915-16; homeotype nos. 6845-49.

Juniperus nevadensis Axelrod
(Plate 4, fig. 22; pl. 17, figs. 20, 24)

Juniperus nevadensis Axelrod, Univ. Calif. Pub. Geol. Sci., vol. 33, p. 278, pl. 12, figs. 9-12; pl. 18, figs. 3-4, 1956.

Wolfe, U.S. Geol. Surv. Prof. Paper 454-N, p. N16, pl. 6, fig. 26, 1964.

Figure 6 illustrates a well-preserved juniper branchlet with three orders of shoots and an attached berry. Characters by which juniper foliage can be separated from that of *Chamaecyparis* are discussed above..

J. nevadensis is very similar to the living *J. osteosperma* (Torr.) Little. The two species share a tendency to decussate phyllotaxis; sharply pointed and consistently overlapping facial leaves; and a scarcity of glands. Some authors (Abrams, 1968, p. 75; Sudworth, 1967, p. 186-189) have noted that *J. osteosperma* is mostly decussate, while *J. californica* Carr., its closest relative, is usually ternate. This is probably most true of the smallest orders of branching, but the difference is only one of degree, since *J. osteosperma* possesses only sporadically the glands which are so ubiquitous on *J. californica* and *J. occidentalis* Hook.

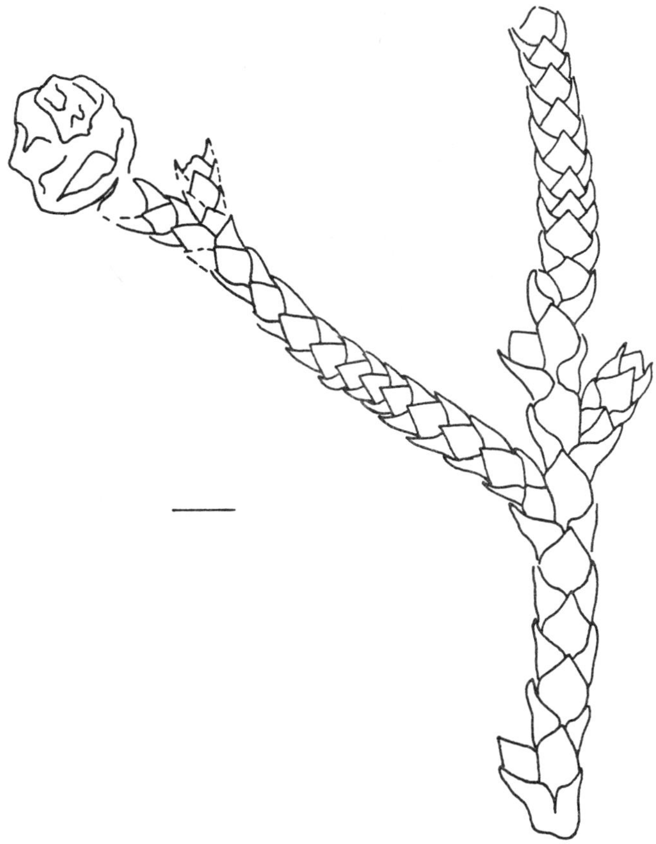

FIGURE 6. Foliage of Juniperus nevadensis Axelrod from the Middlegate flora. Three shoot orders are shown, with a berry on the second-order shoot. Drawing by Steven W. Edwards, 1983. Bar scale = 2 mm.

J. occidentalis has a greater tendency toward overlapping of facial leaves and blunting of facial tips than either californica or osteosperma, and is probably also more consistently ternately whorled than the latter. J. scopulorum Sargent is much more gracile than any of the above species.

At present J. nevadensis can only be distinguished from J. osteosperma in having glands probably more frequently. However, the glands are only occasionally present, rather than ubiquitous as in J. californica. J. nevadensis is almost certainly allied to J. osteosperma and californica rather than to occidentalis or scopulorum.

Occurrence: Middlegate, hypotype no. 6456, homeotype nos. 6457-59; Eastgate, hypotype nos. 6850-51, homeotype nos. 6852-53.

Family TAXODIACEAE

Sequoiadendron chaneyi Axelrod
(Plate 4, figs. 18-20; pl. 18, figs. 1-4;
pl. 19, figs. 1-5)

Sequoiadendron chaneyi Axelrod, Univ. Calif. Pub. Geol. Sci., vol. 33, p. 280, pl. 4, figs. 25-27; pl. 5, figs. 1-3; pl. 17, figs. 1, 4; pl. 25, fig. 4, 1956.
Axelrod, Univ. Calif. Pub. Geol. Sci., vol. 39, p. 228, pl. 43, figs. 3-5, 1962.
Axelrod, Univ. Calif. Pub. Geol. Sci., vol. 51, p. 111, pl. 7, figs. 17-20, 1964.
Axelrod, Missouri Bot. Garden Ann., vol. 63, p. 28, figs. 19, 20, 1976.
Glyptostrobus sp. Wolfe, U.S. Geol. Surv. Prof. Paper 454-N, p. N16, pl. 1, figs. 8, 11, 1964.

Remains of fossil Sierra redwood include several small foliage branchlets in the Middlegate flora, but numerous leafy twigs and branchlets and several cones are in the Eastgate assemblage. Its greater abundance there implies that the trees were closer to the site of plant accumulation, a relation consistent with the more mesic aspect of the Eastgate flora. There is no evidence to show that S. chaneyi differed in any significant way from S. giganteum (Lindley) Buchholtz, which inhabits the western, windward slope of the central and southern Sierra Nevada.

The specimens figured by Wolfe as Glyptostrobus were collected from a stratigraphic horizon above the opalized shales in which most of the Fingerrock flora occurs; this upper flora differs significantly in composition from that recovered at the principal site. Collecting at this upper horizon has yielded several branchlets of Sequoiadendron similar to Wolfe's illustrated specimens; their identity is supported by the presence of *Sequoiadendron* cones in the same beds.

Occurrence: Middlegate, hypotype nos. 6446-48, homeotype nos. 6449-51; Eastgate, hypotype nos. 6821-29, homeotype nos. 6830-41.

Family TYPHACEAE

Typha lesquereuxi Cockerell
(Plate 22, fig. 1)

Typha lesquereuxi Cockerell, Torrey Bot. Club Bull., vol. 33, p. 307, 1906.

Typha latissima Lesquereux, not Al. Braun, Rept. U.S. Geol. Surv. Terr., vol. 8, p. 141, pl. 23, figs. 4, 4a, 1883.

Cattail is common in the Eastgate flora, but less frequent at the Middlegate site. The leaves are linear, ranging up to 2.0 cm wide, and of considerable length, some exceeding 45 cm. They are characterized by numerous longitudinal veins that are interspersed with many fine linear veinlets crossed by numerous short veins normal to the blade. There is no evidence of a midrib in any of the specimens.

The material compares well with leaves of *T. latifolia* Linnaeus, which are typically broader than those of *T. angustifolia* Linnaeus. Their occurrence in a lacustrine section is expectable, judging from the habitats occupied by the modern species.

Occurrence: Middlegate, homeotype nos. 6460-61; Eastgate, hypotype no. 6854, homeotype nos. 6855-59.

Family SPARGANIACEAE

Sparganium nevadense, Axelrod, new species
(Plate 29, figs. 4-5)

Description: Fruiting heads globose, up to 4.5 cm in diameter; each head with numerous, small beaked fruits separated by chaffy, acicular scales up to 0.7-1.0 cm long. Foliage not found.

Discussion: Two fruiting heads of bur-reed in the collection are much larger than any previously described from this region. They are twice as large as specimens from the early Middle Eocene Tipperary flora (MacGinitie, 1974) and fully three times larger than those from the middle Miocene 49-Camp flora of northwestern Nevada (LaMotte, 1936).

Sparganium is represented by 12-15 species of herbaceous perennials that are widely distributed in shallow, aquatic habitats of temperate to cold temperate regions.

Occurrence: Eastgate, holotype no. 7137, paratype no. 7138.

Family SALICACEAE

Populus

Larger samples from old localities as well as collections from new sites make it possible to reevaluate previously described Neogene species of Sect. Tacamahaca in the western states. There are now five readily recognizable species, as follows:

1. Leaves narrowly lanceolate, petioles short

    P. payettensis (Kn.) Axelrod .. cf. P. angustifolia James
    (Idaho City, Thorn Cr.,
    Middlegate, Aldrich Sta.,
    Verdi, Piru Gorge)

2. Leaves ovate-lanceolate, petioles long

    a. Narrowly ovate-lanceolate, base acute to rounded

        P. bonhamii Axelrod .... cf. P. balsamifera Linne.
        (Aldrich Sta., Fallon,
        Middlegate, Eastgate,
        Trapper Cr., Chalk Hills,
        Almaden)

b. Broadly ovate-lanceolate, base rounded to subcordate

P. eotremuloides Knowlton .... cf. P. hastata Dode (P. "trichocarpa" of authors)
(Payette, Succor Cr., Blue Mts., 49-Camp, Eastgate, Middlegate, Wildcat)

3. Leaves broadly ovate

a. Relatively large

P. emersoni Condit .... cf. P. trichocarpa var. trichocarpa T.&G.
(Neroly, Carson Pass)

b. Relatively small

P. verdiana Axelrod .... cf. P. trichocarpa var. trichocarpa T.&G.
(Alturas, Verdi, Deschutes, Sonoma, Piru Gorge)

Populus bonhamii Axelrod, new species
(Plate 5, figs. 4-6; pl. 6, fig. 1;
pl. 20, figs. 1, 4, 7)

Populus eotremuloides Knowlton. Axelrod, Univ. Calif. Pub. Geol. Sci., vol. 33, p. 282, pl. 18, figs. 7, 8, 1956 (Fallon flora).

Axelrod, ibid., p. 282, pl. 26, fig. 5, 1956 (Aldrich Station).

Axelrod, Univ. Calif. Pub. Geol. Sci., vol. 39, p. 229, pl. 45, figs. 1, 3-4, 1962 (Chalk Hills).

Axelrod, Univ. Calif. Pub. Geol. Sci., vol. 51, p. 113, pl. 8, figs. 1-5, 1964 (Trapper Creek).

Populus alexanderi Dorf, Carnegie Inst. Wash. Pub. 412, p. 75, pl. 6, fig. 10 only, 1930 (Alturas).

Axelrod, Univ. Calif. Pub. Geol. Sci., vol. 33, p. 282, pl. 6, fig. 9, 1956 (Aldrich Station).

Description: Leaves narrowly ovate-lanceolate; 8-12 cm long, 3-4 cm broad in lower half of blade; midrib heavy, straight; petiole thick, over 2 cm long; apex narrowly acute, base rounded to obtuse; 8-10 camptodromous secondaries, diverging at moderate angles, the basal pair stronger, extending into broadest part of blade in lower third; thin intersecondaries; tertiaries coarse, irregular; margin with closely spaced, small, crenate teeth or nearly entire; texture firm.

Discussion: This species is characterized by ovate-lanceolate leaves with long tapering tips; the base is chiefly acute to rounded, and the petiole is long. By contrast, the type of *P. eotremuloides* from near Montour (Marsh P.O.), Idaho, has broadly ovate to subcordate leaves. Whereas *P. bonhamii* is most nearly allied to *P. balsamifera* Linnaeus of the eastern United States, *P. eotremuloides* resembles leaves of *P. hastata* Dode (the western *P.* "*trichocarpa*"), distributed from the mountains of California north to Alaska and into the northern and central Rocky Mountains.

The type of *P. trichocarpa* Torrey and Gray, from the Santa Clara River near Ventura, southern California, is a very different taxon, one characterized by broadly ovate leaves. It ranges along the coastal strip and seaward-facing valleys from southern California northward into the San Francisco Bay region. *P. trichocarpa s.s.* is foreshadowed in the fossil record by *P. emersoni* Condit (1938) and by *P. alexanderi* Dorf.

Occurrence: Middlegate, holotype no. 6462, paratype nos. 4272, 6463-65, homeotype nos. 6466-69; Eastgate, hypotype nos. 6860-62, homeotype no. 6863.

Populus cedrusensis Wolfe
(Plate 5, fig. 2; pl. 21, figs. 1-3)

Populus cedrusensis Wolfe, U.S. Geol. Surv. Prof. Paper 454-N, p. N16, pl. 7, figs. 4, 5, 8; pl. 8, fig. 4; text-fig. 7, 1964.

Populus sonorensis Axelrod, Univ. Calif. Pub. Geol. Sci., vol. 33, p. 284, pl. 5, figs. 5, 9-11, 1956.

Populus tremuloides Wolfe, not Michaux, U.S. Geol. Surv. Prof. Paper 454-N, p. N17, pl. 8, fig. 5 only (figs. 6, 7 are *P. pliotremuloides* Axelrod), 1964.

P. cedrusensis is well represented in the Eastgate flora but is rare in the Middlegate. This species has now been recovered from several sites in central to southern Nevada. Its leaves are quite variable in shape from site to site. At Aldrich Station (Axelrod, 1956, pl. 5) they are largely elliptic, yet at Eastgate and Stewart Spring they tend to average more nearly ovate in outline and the petioles are quite long. The material thus appears to resemble leaves produced by two of the living varieties of P. brandegeei Schneider, var. brandegeei Wiggins and var. glabra Wiggins. By contrast, P. sonorensis Axelrod from the Tehachapi, Mount Eden, Anaverde, and Piru Gorge floras in southern California has more consistently ovate to rounded leaves, often subcordate or nearly so.

P. brandegeei is confined now to the mountains of central and southern Baja California and scattered sites in Sonora and adjacent Chihuahua, where it is associated chiefly with sclerophyll vegetation but descends to lower vegetation zones along streamways. In the Miocene of central and southern Nevada it probably occupied streamways in a similar vegetation zone composed of Arbutus, Quercus, and other evergreen dicots.

Occurrence: Middlegate, hypotype no. 6470; Eastgate, hypotype nos. 6864-66, homeotype nos. 6867-70.

Populus eotremuloides Knowlgon

(Plate 6, fig. 5; pl. 20, fig. 6; pl. 21, fig. 6)

Populus eotremuloides Knowlton, U.S. Geol. Surv. 18th Ann. Rept., pt. 3, p. 725, pl. 100, figs. 1, 2; pl. 101, figs. 1, 2, 1898.

LaMotte, Carnegie Inst. Wash. Pub. 455, p. 114, pl. 5, figs. 7, 9, 1936.

Brown, U.S. Geol. Surv. Prof. Paper 186-J, p. 169, pl. 47, fig. 1, 1937.

Chaney and Axelrod, Carnegie Inst. Wash. Pub. 617, p. 151, pl. 17, fig. 4, 1959.

Examination of numerous herbarium sheets of P. hastata Dode (=trichocarpa Torrey and Gray of authors) and P.balsamifera Linnaeus reveals that there are significant differences in the mean leaf shape. The leaves of balsamifera are typically more slender than those of trichocarpa, which tend to be more nearly ovate and some are subcordate (see Sudworth, 1934). Although there is sufficient variation to show that individual leaves cannot be assigned readily to one or the other species, where suites are available they do indicate reasonably well the modern alliance most similar to the fossil. It is on this basis that the rather numerous fossil leaves of taxa of the balsamifera-hastata alliance are separated into two species--eotremuloides and bonhamii.

Wolfe (1964) earlier synonymized all previously described fossil records of P. eotremuloides Knowlton and P. alexanderi Dorf with the living P. trichocarpa Torrey and Gray. This is not acceptable, for it is not demonstrable that all the fossil eotremuloides leaves are identical with trichocarpa. Furthermore, this procedure fails to recognize that there are important differences between the modern and fossil populations, which have significant ecologic implications. This is clearly demonstrated by the fossil records of P. alexanderi in the Deschutes and Verdi floras. They have small ovate leaves like those of P. trichocarpa in the coastal strip from near Santa Cruz southward into southern California. By contrast, trees of hastata ("trichocarpa") at or near the fossil sites produce leaves two to three times larger and are quite different in shape as well (see Axelrod, 1958b, pl. 19-21).

P. eotremuloides Knowlton and P. alexanderi Dorf are therefore reinstated as valid fossil species (Axelrod, 1980, p. 209).

Occurrence: Middlegate, hypotype no. 6471, homeotype nos. 6472-73; Eastgate, hypotype nos. 6871-72.

Populus payettensis (Knowlton) Axelrod
(Plate 5, figs. 1, 3; pl. 21, figs. 4-5;
pl. 22, fig. 3)

Populus payettensis (Knowlton) Axelrod, Carnegie Inst. Wash. Pub. 553, p. 253, 1944.

Rhus payettensis Knowlton, U.S. Geol. Surv. 18th Ann. Rept., pt. 3, p. 733, pl. 101, figs. 6, 7, 1898.

Celastrus lindgreni Knowlton, ibid., p. 732, pl. 99, fig. 13; pl. 100, fig. 6, 1898.

Populus dolichophylla Smith, Amer. Midland Natur., vol. 25, p. 494, pl. 3, fig. 5; pl. 4, figs. 1, 2, 4, 1941.

Populus payettensis (Knowlton) Axelrod. Axelrod, Carnegie Inst. Wash. Pub. 553, p. 281, pl. 48, fig. 4, 1944.

Axelrod, Carnegie Inst. Wash. Pub. 590, p. 200, pl. 4, figs. 4, 5, 1950.

Axelrod, Univ. Calif. Pub. Geol. Sci., p. 283, pl. 5, figs. 4, 6-8, 12; pl. 13, fig. 5; pl. 26, fig. 3, 1956.

Several slender, lanceolate leaves in both floras are similar to those produced by the living P. angustifolia James of the western United States. The leaf margins are finely crenate and gland-tipped, the petioles are thick and short, and the short, basal primaries are plinerved.

These are not the sucker-shoot leaves of the fossil P. eotremuloides. Although the living P. hastata (=P. trichocarpa), which is allied to eotremuloides, produces leaves somewhat similar to those of P. payettensis, they can be separated because the latter has typically very short petioles. It is apparent that the type specimens, described originally by Knowlton as Rhus nevadensis and Celastrus lindgreni, do not represent sucker-shoot leaves of P. eotremuloides. The types come from Idaho City and so far as is known are not associated with leaves of P. eotremuloides, which occur at Cartwright Ranch and near Montour; they are much larger and commonly nearly cordate. It is also noted that the Idaho City locality is only a few miles from the

Thorn Creek flora, in which leaves of P. payettensis described as P. dolichophylla Smith) are very abundant. A detailed geologic study may well show that these basins were formerly connected and have since been isolated by faulting.

Occurrence: Middlegate, hypotype nos. 6475-76, homeotype nos. 6477-79; Eastgate, hypotype nos. 6873-75.

Populus pliotremuloides Axelrod
(Plate 20, figs. 2-3, 5)

Populus pliotremuloides Axelrod, Carnegie Inst. Wash. Pub. 476, p. 169, pl. 4, figs. 1-3, 1937.
Axelrod, Carnegie Inst. Wash. Pub. 590, p. 53, pl. 2, fig. 4, 1950.
Axelrod, Univ. Calif. Pub. Geol. Sci., vol. 33, p. 284, pl. 25, fig. 10, 1956.
Axelrod, Univ. Calif. Pub. Geol. Sci., vol. 34, p. 128, pl. 22, figs. 5-8, 1958.
Axelrod, Univ. Calif. Pub. Geol. Sci., vol. 39, p. 230, pl. 44, figs. 4, 5, 1962.
Populus tremuloides Wolfe, not Michaux, U.S. Geol. Surv. Prof. Paper 454-N, p. N17, pl. 8, figs. 6, 7, 1964.

Only two additional specimens have been recovered from the Middlegate flora, but they are more abundant in the Eastgate assemblage. The difference in representation reflects the local setting at these sites, with the species confined to higher levels in the drier, warmer Middlegate area and hence its poorer representation there.

As discussed for the records of P. eotremuloides, Wolfe's assignment of the fossil species to the living P. tremuloides is rejected for the same reasons. Furthermore, his assertion that P. pliotremuloides is known only from rocks of Barstovian and later age is belied by its occurrence in the Eastgate and Middlegate floras, both dated at 18.5 m.y., and by its record at Creede, Colorado, where the well-dated section is 26.5 m.y. old.

Leaves of *P. pliotremuloides* are consistently smaller than those of the allied *P. voyana* Chaney and Axelrod from the Columbia Plateau region (Chaney and Axelrod, 1959, p. 152). A similar species described from the Alaskan Miocene, *P. kenaiana* Wolfe (1966; Wolfe and Tanai, 1980) differs in no respects from *P. voyana*, which has name priority.

Occurrence: Middlegate, hypotype no. 4277, homeotype no. 4278; Eastgate, hypotype nos. 6876-78, homeotype nos. 6879-80.

Salix desatoyana Axelrod, new species
(Plate 22, figs. 2, 6, 7)

Description: Leaves linear-lanceolate; base acute, the tip long acuminate; leaves 7.0-8.0 (estim.) cm long and 0.7 cm broad; petiole scarcely 1 cm long, thick; numerous secondaries diverge at moderate angles and then curve sharply upward subparallel to the margin, forming a marginal vein from which thin secondaries supply marginal teeth; numerous short intersecondaries merge into a coarse, open polygonal mesh; margin with sharp serrate teeth; texture firm.

Discussion: Two broken leaves in the Eastgate assemblage and one in the Buffalo Canyon flora are similar to those produced by *Salix nigra* Marshall. This is a widespread tree along watercourses in the eastern United States, reaching westward into the High Plains.

The only previously described fossil willow that shows any relationship to *S. desatoyana* is *S. truckeana* Chaney (1944, pl. 52, figs. 2-6), a broader-leaved black willow allied to *S. gooddingii* Ball, distributed from the central valley of California southeastward through southern Arizona and New Mexico to the Rio Grande valley in western Texas and extending into northern Mexico (see Sudworth, 1934, p. 58).

Occurrence: Buffalo Canyon, holotype no. 6881; Eastgate, paratype nos. 6882-83.

Salix owyheeana Chaney and Axelrod

Salix owyheeana Chaney and Axelrod, Carnegie Inst. Wash. Pub. 617, p. 154, pl. 18, fig. 2, 1959.
Axelrod, Univ. Calif. Pub. Geol. Sci., vol. 39, p. 231, pl. 46, fig. 7 only (not fig. 6, which is *Salix boisiensis* Smith), 1962.
Arbutus prexalapensis Axelrod, Univ. Calif. Pub. Geol. Sci., vol. 33, pl. 32, fig. 3 only (figs. 1, 2, 4 remain *Arbutus*), 1956.

Examination of the above-cited *Arbutus* leaf under cross-lighting that falls parallel to the leaf blade indicates that creases in the rock matrix that simulate the venation of madrone led to the misidentification: it is *Salix*. The leaves of this willow show relationship to those produced by *S. hookeriana* Barratt of the coastal strip from northern California northward. *S. bebbiana* Sargent, a widely distributed species in the eastern United States and Rocky Mountains, also has similar leaves, though most average somewhat smaller than the fossil.

Occurrence: Middlegate, hypotype no. 4406.

Salix pelviga Wolfe
(Plate 6, figs. 6-7; pl. 22, figs. 4, 8, 12)

Salix pelviga Wolfe, U.S. Geol. Surv. Prof. Paper 454-N, p. N18, pl. 8, figs. 1, 2, 8; text-fig. 8, 1964.
Salix hesperia (Knowlton) Condit. Axelrod, Univ. Calif. Pub. Geol. Sci., vol. 33, p. 285, pl. 25, fig. 11, 1956.
Salix truckeana Chaney. Axelrod, ibid., p. 286, pl. 26, fig. 2, 1956.

There is a large suite of leaves representing this species in the Eastgate flora, and several additional specimens have been recovered from the Middlegate site. They show that the variation of the species includes that of the two species listed above for the Middlegate flora.

The leaves of this willow compare favorably with the variation displayed by those of _S_. _melanopsis_ Nuttall, a shrub or small tree widely distributed in the mountains of central California northward to British Columbia and east to the Rocky Mountains.

Occurrence: Middlegate, hypotype nos. 6480-81, homeotype nos. 6482-85; Eastgate, hypotype nos. 6884-86, homeotype nos. 6887-90.

Salix storeyana Axelrod
(Plate 6, figs. 3-4; pl. 23, figs. 7-8)

Salix storeyana Axelrod, Univ. Calif. Pub. Geol. Sci., vol. 121, p. 211, 1980.

Salix knowltoni Berry. Axelrod, Univ. Calif. Pub. Geol. Sci., vol. 33, p. 285, pl. 7, figs. 1, 2; pl. 13, figs. 6, 7; pl. 19, figs. 1, 2; pl. 25, figs. 12, 13, 1956.

Axelrod, Univ. Calif. Pub. Geol. Sci., vol. 39, p. 231, pl. 45, fig. 2.; pl. 26, figs. 1-3, 5, 1962.

There are a dozen or so leaves of _S_. _storeyana_ in the Middlegate flora and three are in the Eastgate assemblage. They are readily recognized by their entire margin and the gently looping, nearly subparallel secondaries. The leaves figured for the Aldrich Station flora are small and obovate and thus somewhat abnormal for the species, but otherwise have all its characters. Similar leaves are produced by _S_. _lemmonii_ Bebb, which inhabits the mixed-conifer and fir forests of the Sierra Nevada slope, ranging north into Oregon and southward into the mountains of southern California.

Occurrence: Middlegate, hypotype nos. 6486-87, homeotype nos. 6488-96; Eastgate, hypotype nos. 6891-92, homeotype nos. 6893-95.

Salix venosiuscula Smith
(Plate 6, fig. 2; pl. 22, fig. 11)

Salix venosiuscula Smith, Amer. Midland Natur., vol. 25, p. 500, pl. 3, fig. 7; pl. 4, fig. 6, 1941.
Chaney and Axelrod, Carnegie Inst. Wash. Pub. 671, p. 154, 1959 (see synonymy and discussion).

A single leaf in the Eastgate and two in the Middlegate flora represent this species. They are similar to leaves produced by S. caudata (Nuttall) Heller, which ranges from northeastern California north into Canada east of the Cascades and into the Rocky Mountains. As reviewed elsewhere (Chaney and Axelrod, 1959, p. 154), S. venosiuscula has been recorded in the Thorn Creek, Hog Creek, Alturas, and Deschutes floras of Idaho, eastern Oregon, and northeastern California.

Occurrence: Middlegate, hypotype no. 6497, homeotype no. 6498; Eastgate, hypotype no. 6896.

Salix wildcatensis Axelrod

Salix wildcatensis Axelrod, Carnegie Inst. Wash. Pub. 553, p. 132, 1944 (see synonymy).
Chaney, Carnegie Inst. Wash. Pub. 553, p. 341, pl. 58, fig. 2; pl. 59, figs. 1-4, 1944.
Axelrod, Carnegie Inst. Wash. Pub. 590, p. 204, pl. 4, fig. 9; pl. 5, fig. 8, 1950.
Axelrod, Univ. Calif. Pub. Geol. Sci., vol. 33, p. 287, pl. 25, fig. 14, 1956.

The single specimen recorded earlier from the Middlegate flora remains the sole record of this distinctive willow at this site. Among modern species, leaves of S. lasiolepis Bentham of western California and southern Arizona are quite similar.

Occurrence: Middlegate, hypotype no. 4288.

Family JUGLANDACEAE

Juglans nevadensis Berry
(Plate 9, figs. 4-5)

Juglans nevadensis Berry, Wash. Acad. Sci. Jour., vol. 18, p. 158, fig. 1, 1928.

Juglans beaumontii Axelrod (in part), Carnegie Inst. Wash. Pub. 476, p. 171, pl. 4, fig. 12 only (not fig. 11, which remains *J. beaumontii*), 1937.

Axelrod, Carnegie Inst. Wash. Pub. 590, p. 103, 1950 (see discussion).

A leaflet and its counterpart in the Middlegate flora are referred to this species, which is represented by numerous walnuts in the Mount Eden collection at the Los Angeles Natural History Museum; by well-preserved nuts in the middle Miocene Puente flora near Puddingstone Dam, southern California (U.C. Mus. Paleo.); and by the type specimen from the middle Truckee Formation of Hemphillian age (5-6 m.y.).

Supplementary description: Leaflets oblong-lanceolate, base asymmetrical and rounded, with a short petiolule; tip evidently acute; midrib firm; leaflets 4.0-4.5 cm long (estimated) and 1.6 cm broad in lower third of blade, its widest part; alternate looping secondaries subcamptodrome, either supplying the serrate to blunt teeth or forking near margin to enter the teeth; tertiaries irregular, forming a relatively open pattern; quaternary mesh open and irregular; texture medium.

Discussion: These leaflets are similar to those produced by *J. californica* S. Watson, a small tree widely distributed in the coastal valleys and hills of southern California from near Santa Barbara southward into Orange County (see Griffin and Critchfield, 1972).

The occurrence of *J. nevadensis* in the Middlegate flora is not unexpected, for a number of Madrean taxa--*Ceanothus*, *Cercocarpus*, *Fraxinus*, *Lyonothamnus*, *Platanus*, *Quercus*, *Robinia*--were recorded there previously (Axelrod, 1956).

Occurrence: Middlegate, hypotype nos. 6499a-b.

Family BETULACEAE

Identification of fossil leaves of the genera in this family is not always a simple task. Diagnostic features are often not visible, owing to the fineness of detail required to separate certain taxa. Furthermore, the morphologic features that separate some taxa are not clearly understood. For example, Klucking (1959) studied the Betulaceae and concluded that _Alnus_ and _Betula_ could be separated on the basis of secondary or tertiary venation: the ends of the veins curve distally in _Betula_, basally in _Alnus_, a criterion also accepted by Wolfe (1964, 1966). This is indeed true for _some_ species, but is _not_ diagnostic of the genera. Leaves of a number of alder species show no preference for the secondaries to curve at all, as in _A_. _acuminata_, _fruticosa_, _glutinosa_, _jorullensis_, _oregona_, _rhombifolia_, _rugosa_, and _tenuifolia_; they are craspedodrome or semicraspedodrome. Other species of _Alnus_ have secondaries and distal tertiaries that curve apically (not basally), as in _A_. _acuminata_, _firma_, _incana_, _japonica_, _matsumurae_, _obtusata_, _pendula_, and _sieboldiana_. In some species, different leaves on the same twig have secondaries that curve in both directions, as in _A_. _acuminata_, _cremastogyne_, _crispa_, _lanata_, _nagurae_, _maximowiczii_, and _sinuata_. And some alders have camptodromous secondaries, or nearly so, as _A_. _cremastogyne_, _cordata_, _faurei_, _nepalensis_, _nitida_, and _trabeculosa_.

In the genus _Betula_, the following have secondaries and branching distal tertiaries that do curve apically: _B_. _corylifolia_, _cylindrostachya_, _ermanii_, _japonica_, _lenta_, _luminifera_, _lutea_, _maximowiczii_, _nigra_, _occidentalis_, _populifolia_, _pubescens_, _pumila_, _utilis_, and _verrucosa_. However, _B_. _alnoides_ and _papyrifera_ curve both distally and basally, and _insignis_ and _occidentalis_ are craspedodrome or semicraspedodrome and loop in both directions.

These data indicate that while a fossil leaf with secondaries or tertiaries that loop distally at the margin may well be _Betula_, it may also be _Alnus_, as indicated by the listing above. Clearly, shape, percurrency, and other features must all be considered if errors in generic assignment are to be reduced to a minimum. A further complication is the fact that fossil species often are represented by only one or two specimens, so that the variation of the population, or the species, is not known. Since _all species_ with non-entire leaves show some diversity in leaf form and marginal characters, it is not always possible to separate taxa on minor characters of marginal venation, shape of teeth, or the nature of the apex or base.

Studies must be based on variation in leaves shown by the species over its range, not by reference to one or two cleared leaves that display leaf architecture, but not the diversity encountered in the species. This practice can and has led to errors in assignment of taxa. For example, Wolfe (1966, and Wolfe and Tanai, 1980) identified 10 alder species in the Neogene of Alaska (see key, Wolfe, 1966, p. B15), separating _A. cappsii_ (Hollick) Wolfe from _A. corylina_ Knowlton and Cockerell on the basis of whether the teeth are sharp or rounded. However, leaves of _A. tenuifolia_ Nuttall--placed by some in _A. incana_ ssp. _tenuifolia_ (Nutt.) Breitung--produce leaves with _both_ kinds of teeth, as can readily be seen in the herbarium; the fossil species _cappsii_ and _corylina_ are therefore conspecific. According to Wolfe's key to identification, _A. cappsii_ has a decurrent leaf base, but this is not evident in the illustrations, including the types (Hollick, 1936) which are misnumbered in Wolfe (1966, p. B19) as pl. 49, fig. 3, instead of pl. 71, fig. 3. Furthermore, Wolfe's key states that the teeth are sharp, yet pl. 6, figs. 1, 4, show rounded lobations, whereas in Wolfe and Tanai (1980), pl. 8, fig. 5, illustrates that they are acute; thus, the figures do not match the key to identification.

As for _A. schmidtae_ Wolfe and _A. barnesii_ Wolfe, they are separated on the basis of whether the tips are abruptly acute or acuminate. The illustrated leaf of _A. schmidtae_ does not have an apex (Wolfe, 1966, pl. 5, fig. 1). The overall similarity of _schmidtae_ and _barnesii_ is so great that they appear to be the same species, if one admits that variation does occur in leaves of alders and other taxa. Fossil leaves of both these species can also be matched by leaf variation in *A. tenuifolia*. Although Wolfe separated *A. schmidtae* and *barnesii* from *A. corylina* and *cappsii* on the basis of whether there are 7-8 or 5-6 nervilles per centimeter, this variation may only express the difference in sun and shade leaves. In my opinion, all four of these species--*corylina*, *cappsii*, *schmidtae*, and *barnesii*--represent the variation of one fossil plant, *A. corylina* Knowlton and Cockerell. The latter three are therefore reduced to synonymy.

Alnus harneyana Chaney and Axelrod
(Plate 7, fig. 5)

Alnus harneyana Chaney and Axelrod, Carnegie Inst. Wash. Pub. 617, p. 158, pl. 21, figs. 3-9, 1959.

A single, poorly-preserved, incomplete leaf in the Middlegate flora is the sole record of the species in this area. The leaf compares favorably with those produced by *A. tenuifolia* Nuttall, a widely distributed streamside shrub in the forests of the western United States, ranging north into the Yukon.

As noted earlier (Chaney and Axelrod, 1959, p. 158), leaves of *A. corylina* Knowlton and Cockerell from Alaska appear to be similar, as do other Alaskan species described by Wolfe (1966) and Wolfe and Tanai (1980). Since the Alaskan species have been greatly overspeciated (see above), it is not easy to evaluate their relations, especially since their relationship to living taxa, and the variation of the fossil species themselves, have not been indicated.

Alder leaves of any species do vary to some degree in shape, lobation, and number and shape of teeth, the latter characters that Wolfe selected as diagnostic of the fossil species. Furthermore, it must be noted that (1) the older Oligocene-early Miocene species would be expected to show greater variation in leaf-form venation and marginal features than modern derivative species; and (2) these older taxa were no doubt hybridizing to some extent, thus adding further to confusion in their identification--factors not considered by Wolfe. Until the suites of 10 Miocene Alaskan alders can all be assembled and restudied together with adequate herbarium specimens, it seems best to retain the western Miocene species separate from the Alaskan material.

Occurrence: Middlegate, hypotype no. 6500.

Alnus largei (Knowlton) Wolfe
(Plate 7, fig. 1)

Alnus largei (Knowlton) Wolfe, U.S. Geol. Surv. Prof. Paper 398-B, p. B16, pl. 7, fig. 5; fig. 2, 1966 (see synonymy).

A single leaf and its counterpart in the Middlegate flora are referred to this species. Wolfe states that one of its key characters is the fact that the leaves are without lobations (teeth in groups), yet several of those that he has listed in the synonymy definitely have lobations. Specimens of the living east Asian species, *A. cremastogyne* Burkill (U.C. Sheet 224253) and *A. trabeculosa* Hand.-Mzt., which resemble the fossil, have leaves of both sorts. These alders are found in the temperate forests of east-central China, where many other taxa have allied species in the Tertiary of the western United States. *A. largei* is well represented in the late Eocene Elko flora, at the site of the old coal mines directly south of town.

Occurrence: Middlegate, hypotype no. 6501.

Betula thor Knowlton
(Plate 23, figs. 3, 6)

Betula thor Knowlton, U.S. Geol. Surv. Prof. Paper 140, p. 35, pl. 17, fig. 3, 1926.

Chaney and Axelrod, Carnegie Inst. Wash. Pub. 617, p. 160, pl. 23, figs. 2-6, 1959 (see synonymy).

Axelrod, Univ. Calif. Pub. Geol. Sci., vol. 33, p. 289, pl. 26, fig. 1, 1956.

Axelrod, Univ. Calif. Pub. Geol. Sci., vol. 39, p. 232, pl. 47, figs. 1-4, 1962.

Axelrod, Univ. Calif. Pub. Geol. Sci., vol. 151, p. 117, pl. 10, figs. 1, 2, 7, 1964.

Leaves and cones in the Eastgate flora and several fragmentary leaves in the Middlegate flora represent this species. They are birch on the basis of general shape, the relatively simple marginal teeth--alders of similar shape have more numerous serrations, as in _A_. _sinuata_ (Regel) Rydberg--and forward-curving secondaries where they enter the primary marginal teeth. The fossil leaves are well matched by those of paper birch, _B_. _papyrifera_ Marshall, that ranges from the eastern United States across Canada and southward into Washington, northeastern Oregon, and the Idaho panhandle.

Occurrence: Middlegate, homeotype nos. 6502-03; Eastgate, hypotype nos. 6899-6900, homeotype nos. 6903-14.

Betula _vera_ Brown

(Plate 7, figs. 2-4; pl. 23, fig. 9)

Betula _vera_ Brown, U.S. Geol. Surv. Prof. Paper 186, p. 171, pl. 48, figs. 7-11, 1937.

Axelrod, Univ. Calif. Pub. Geol. Sci., vol. 33, p. 290, pl. 26, fig. 6, 1956.

Chaney and Axelrod, Carnegie Inst. Wash. Pub. 617, p. 161, pl. 23, fig. 8, 1959.

Several additional ovate leaves of this species have been recovered from the Middlegate site. They are acute at the tip, have a rounded base, and the margin is doubly serrate with rather uniform acute teeth. The secondaries are evenly spaced and only slightly curved.

As noted by Brown (1937), the leaves of _B_. _vera_ are similar to those produced by the yellow birch, _B_. _lutea_ Michaux, of the eastern United States, which inhabits the area from the Great Lakes eastward into Maine and Nova Scotia, reaching southward into the Appalachian Mountains of the Carolinas.

Occurrence: Middlegate, hypotype nos. 6504-06, homeotype nos. 6507-10; Eastgate, hypotype no. 6917, homeotype no. 6918.

Family FAGACEAE

Chrysolepis convexa (Lesq.) Axelrod, new comb.

(Plate 9, figs. 1-3; pl. 23, figs. 1-2)

Castanopsis convexa (Lesquereux) Brooks, Carnegie Mus. Ann., vol. 24, p. 288, 1936.

Quercus convexa Lesquereux, Harvard Mus. Comp. Zool. Mem., vol. 6, no. 2, p. 4, pl. 1, figs. 13-17, 1878.

Condit, Carnegie Inst. Wash. Pub. 553, p. 76, pl. 14, figs. 2, 3, 5, 6, 1944.

Brooks transferred the type specimens of Q. convexa to Castanopsis, and figured many specimens from Succor Creek as Castanopsis. However, Graham (1965, p. 78) reports that no pollen of Castanopsis was found in the Succor Creek flora, even though fully 40,000 grains were examined. A more recent study (Taggart and Cross, 1980) also has not recognized Castanopsis pollen in the Succor Creek flora. Since the Succor Creek foliar specimens rarely preserve the finer details of venation, most of the leaves identified as C. convexa at Succor Creek must remain open to question for the moment. However, some of them (U.C. Mus. Paleo.) are sufficiently well preserved to show that they are indeed Quercus, and are referred tentatively to Q. dayana Knowlton.

Comparison of the type specimens and supplementary material of Q. convexa from Table Mountain does indeed show that the leaves are Chrysolepis (=Castanopsis). They differ from the somewhat similar leaves of certain species of Quercus (e.g., Q. virginiana) in the fourth- and fifth-order nervation, which is composed of finer veins and hence does not appear so dense as in Quercus. The specimens compare favorably with the variation in size and shape shown by leaves of C. (Castanopsis) sempervirens (Kellogg) Hjelmquist, a common shrub in the middle and upper parts of the Sierran mixed-conifer forest and the fir forest belt as well. Its occurrence at lower levels in the Miocene forest zones, where it was adjacent to broadleaved sclerophyll woodland, is attributed to summer rain that enabled the species to live there under conditions in which severe drought stress was absent.

Occurrence: Middlegate, hypotype nos. 6511-13, homeotype nos. 6514-18; Eastgate, hypotype nos. 6919-20, homeotype nos. 6921-31.

Chrysolepis sonomensis (Axelrod) Axelrod, new comb.
(Plate 10, figs. 1-3)

Castanopsis sonomensis Axelrod, Carnegie Inst. Wash. Pub. 553, p. 196, 1944 (see synonymy and discussion).

Axelrod, Carnegie Inst. Wash. Pub. 590, p. 57, pl. 2, fig. 13, 1950.

Axelrod, Univ. Calif. Pub. Geol. Sci., vol. 39, p. 233, pl. 48, fig. 1, 1962.

Several large, lanceolate Middlegate leaves of C. sonomensis have acute, sharp, pointed tips and rounded to acute bases. The leaves are entire, have a relatively long, stout petiole, numerous subparallel secondaries that are nearly camptodrome, and a fine tertiary and quadrangular mesh. The specimens are similar to leaves produced by the living C. (Castanopsis) chrysophylla (Douglas) Hjelmquist, a large tree distributed from the central Coast Ranges of California northward into coastal Oregon. A relict patch in Placer County, California, east of Georgetown, in the lower Sierran mixed-conifer forest, is adjacent to sclerophyll vegetation dominated by Arbutus, Lithocarpus, and Quercus chrysolepis. C. sonomensis probably had a similar occurrence in the Middlegate flora.

Occurrence: Middlegate, hypotype nos. 6519-21.

<u>Lithocarpus</u> <u>nevadensis</u> Axelrod, new species
(Plate 8, figs. 8-10; pl. 24, figs. 1-7)

<u>Quercus</u> <u>simulata</u> Knowlton. Condit, Carnegie Inst. Wash. Pub. 553, p. 45, pl. 5, fig. 3, 1944.
Axelrod, Univ. Calif. Pub. Geol. Sci., vol. 33, p. 291, pl. 13, fig. 12; pl. 20, fig. 6; pl. 27, figs. 1-4, 1956.
Axelrod, Univ. Calif. Pub. Geol. Sci., vol. 39, p. 223, pl. 48, fig. 3, 1962.

<u>Description</u>: Leaves lanceolate, 7-11 cm long, 1.5-3.0 cm broad; apex acute, base acute to rounded; midrib strong and thick, thinning above; petiole thick and heavy, 1.0-1.5 cm or more long; 11-16 alternate thick secondaries, occasional intersecondaries; craspedrodrome where teeth present, otherwise poorly camptodromous; tertiaries form a heavy, coarse network nearly normal to secondaries; finer nervation dense and heavy; margin entire or with small, coarse teeth, chiefly in distal portion; texture heavy.

<u>Discussion</u>: The leaves of <u>L</u>. <u>nevadensis</u> differ from those of <u>L</u>. <u>klamathensis</u> (MacGinitie) Axelrod in having a consistently more slender, lanceolate shape. The difference is especially apparent when large suites of fossil leaves are available for study, as in the present floras.

The venation in <u>Lithocarpus</u> leaves is distinctive in that the nearly parallel, craspedodrome secondaries waver in their outer portion and enter the teeth directly. Also, the tertiary venation is not as strongly percurrent as in Asian species of the genus. The problem of separating some of these serrate leaves from those of <u>Q</u>. <u>simulata</u> has not been clearly solved. As noted earlier, serrations in leaves of <u>simulata</u> may extend down to near the middle of the leaf, rarely lower. The problem is that leaves of <u>Lithocarpus</u> may also have only a few teeth distally, or none--being entire. Most of the leaves of the slender form (<u>lanceolata</u>) of <u>Lithocarpus</u> have bluntly rounded to acute tips and for the most part are not as sharply attenuated as those of <u>Q</u>. <u>simulata</u>.

Certainly, the specimens of Q. simulata at the type locality (Cartwright Ranch) are markedly acuminate, as judged from the large collections made there recently by Patrick Fields. This is true not only of the serrate leaves but of the entire ones as well.

The leaves of Lithocarpus are abundant in both floras of the Middlegate basin. Specimens in the Eastgate flora include remains of the distinctive acorn cup with spreading hair-like scales that also typifies the genus. Leaves from the two sites differ somewhat in size, with those from Eastgate tending to be larger. This presumably reflects the more mesic habitat at Eastgate, as indicated by the greater representation of mesic taxa there, notably Abies, Betula, Tsuga, Sequoiadendron, and others.

Leaves of L. densiflorus show an important change in shape in the interior, where they become lanceolate, a form recognized earlier by Jepson (1910) as forma lanceolata. The fossils closely resemble this ecotype.

Occurrence: Middlegate, holotype no. 4325, paratype nos. 4322-24, 4326-29, hypotype nos. 6522-24, homeotype nos. 6525-36; Eastgate, hypotype nos. 6932-38, homeotype nos. 6939-41.

Quercus hannibali Dorf
(Plate 8, figs. 1-7; pl. 25, figs. 1-7)

Quercus hannibali Dorf, Carnegie Inst. Wash. Pub. 412, p. 86, pl. 8, fig. 9 only, 1930.

Chaney and Axelrod, Carnegie Inst. Wash. Pub. 617, p. 168, pl. 24, fig. 2; pl. 25, figs. 11-13, 1959 (see synonymy).

Axelrod, Univ. Calif. Pub. Geol. Sci., vol. 121, p. 56, pl. 7, figs. 1-5, 1980.

Quercus chrysolepis Wolfe, not Liebmann, U.S. Geol. Surv. Prof. Paper 454-N, pl. 2, figs. 1-10, 14; pl. 9, figs. 2, 3, 5-7, 12, 16, 1964 (and all items in synonymy).

Leaves of Q. hannibali are abundant in both floras, but only occasional acorn cups have been recovered. This is not surprising, for their area-weight ratio would make unlikely their transport far from the lake margin, where the species presumably occupied warmer sites, for the most part.

Relationship to the living Q. chrysolepis Liebmann is apparent. This species is a regular member of the broad-leaved sclerophyll forest, reaching up into the lower part of the mixed-conifer forest in warmer sites, chiefly.

Occurrence: Middlegate, hypotype nos. 6537-44, homeotype nos. 6545-61; Eastgate, hypotype nos. 6942-48, homeotype nos. 6949-60.

Quercus shrevoides Axelrod, new species
(Plate 9, figs. 6-10; pl. 25, figs. 8-10; pl. 29, fig. 11)

Quercus wislizenoides Axelrod, Carnegie Inst. Wash. Pub. 553, p. 46, pl. 5, figs. 4, 5, 1944.
Axelrod, Univ. Calif. Pub. Geol. Sci., vol. 33, p. 291, pl. 14, figs. 1, 2; pl. 20, figs. 4, 5, 8; pl. 27, figs. 5-8, 1956.
Quercus cedrusensis Wolfe, U.S. Geol. Surv. Prof. Paper 454-N, p. N21, pl. 9, fig. 15, 1964.

Description: Leaves lanceolate to broadly elliptic, 4-9 cm long, 2.5-4.5 cm broad; apex acute to blunt, with sharp teeth, base rounded to acute; petiole thick and heavy, 5-15 mm long; 5-8 alternate secondaries, diverging at medium-high angles, wavering in outer part, branching to enter marginal teeth where present, otherwise semicamptodromous; coarse, irregular tertiary network; margin entire or with sharp teeth or slightly lobed; texture heavy.

Discussion: Leaves of Q. shrevoides are well represented in both the Middlegate and Eastgate floras. Their relatively large size, as well as the venation and marginal features, compare closely with leaves produced by Q. shrevei C. H. Muller. This large forest tree is a codominant of the sclerophyll forest of the central Coast Ranges, especially in the Santa Lucia Mountains (pl. 2, fig. 1). The allied *Q. wislizenii*, which contributes to the interior pine-oak woodland dominated also by *Q. douglasii* and *Pinus sabiniana*, produces leaves that are typically smaller, regularly crisped, and more sharply serrated.

Q. cedrusensis Wolfe (1964, pl. 9, fig. 15) differs in no way from leaves of the modern *Q. shrevei*. Although it has priority, this name is hereby rejected because of confusion with the living species *Q. cedrosensis* C. H. Muller.

Occurrence: Middlegate, holotype no. 6570, paratype nos. 6571-90, 4331-4334; Eastgate, hypotype nos. 6970-72, 7139, homeotype nos. 6973-76; Chloropagus, hypotype nos. 4232-33; Fallon, hypotype nos. 3955-56, 2077; Remington Hill, hypotype nos. 2391-92.

Quercus simulata Knowlton

(Plate 10, figs. 4-6; pl. 26, figs. 6-9)

Quercus simulata Knowlton, U.S. Geol. Surv. 18th Ann. Rept., pt. 3, p. 728, pl. 101, fig. 4; pl. 102, figs. 1, 2, 1898.

Chaney and Axelrod, Carnegie Inst. Wash. Pub. 617, p. 171, pl. 30, figs. 2, 3, 5-8; pl. 31, figs. 1-4, 1959 (see synonymy).

Judging from the finer nervation, *Q. simulata* may be an extinct member of the *Q. chrysolepis* alliance. Although it has been compared with *Q. myrsinaefolia* Blume, the tertiaries of the leaves of that species have a relatively strong cross-percurrent arrangement. And those of *Q. stenophylla*, which has also been compared with *sinuata*, have an even stronger cross-percurrent tertiary venation than those of *simulata*.

As noted above, these leaves can generally be separated from the rather similar leaves of *Lithocarpus* by their more slender, lanceolate outline; the tips are more attenuated, often being long-acuminate, and the tertiaries are not prominently percurrent. In the serrate leaves, the secondaries of *Lithocarpus* waver in their outer portion, whereas in serrate *Q. simulata* they are strongly craspedodrome and firmer.

It is possible that the specimens referred to *Q. simulata* Knowlton differ sufficiently from the type material to warrant their recognition as a new species.

Occurrence: Middlegate, hypotype nos. 6562-64, homeotype nos. 6565-69; Eastgate, hypotype nos. 6961-64, homeotype nos. 6965-69.

Family BERBERIDACEAE

Mahonia

Mahonia was already a distinct genus in the late Cretaceous (Brown, 1950), and is associated with *Berberis* (*sensu stricto*) in the late Eocene (Bull Run flora, Nevada) and late Oligocene (Creede flora, Colorado). *Mahonia* differs from *Berberis* in several features, as outlined by Fedde (1901) and Ahrendt (1961). *Mahonia* is evergreen, *Berberis* deciduous. *Mahonia* has pinnate leaves with numerous, very coriaceous, usually quite spiny leaflets, whereas *Berberis* has small leaves of thinner texture and the margins are entire, serrate, or sparsely toothed. Other features that serve to separate these genera are not preserved as fossil.

Identification of fossil species of *Mahonia* is not an easy task, because leaflets of species overlap in characters and they vary in shape and marginal features, depending on position on the rachis. Furthermore, fossil leaflets similar to those of living species may be difficult (or impossible) to separate, because the modern species often are based on characters (color of leaves, pubescence, etc.) that are not preserved (see Ahrendt, 1961). When a large suite of fossil specimens (typically individual leaflets) from one locality is available, it is usually possible to recognize a natural entity, as compared with a sample of only one or two specimens. In the present case, relatively large suites of *Mahonia* leaflets in the Eastgate flora aid in clarifying the relations of its species to those in fossil floras in the bordering area to the north.

The following treatment resulted from consultation with Howard Schorn and Patrick Field, using all material in the

Museum of Paleontology and the Herbarium, University of California, Berkeley. It is emphasized that the grouping of previously described and figured species into the following taxa is based on the principle that, even though the leaflets differ individually to a considerable degree (some paleobotanists might well name them distinct species), all of them can be matched by the variations in leaflets of a single species, or closely allied species, over the area of its distribution. This is consistent with the fact that the fossil species, like the modern, have been collected from widely scattered sites and show similar variation in leaflet characteristics.

Mahonia macginitiei Axelrod, new species
(Plate 11, figs. 2, 4, 9; pl. 27, fig. 7)

Odostemon hollicki Dorf. MacGinitie, Carnegie Inst. Wash. Pub. 614, p. 55, pl. 7, figs. 1, 3, 5, 1933 (Trout Cr.).

Arnold, Univ. Mich. Mus. Paleo. Contrib., vol. 5(4), p. 61, pl. 2, figs. 3, 4, 7 (Succor Cr.), figs. 5, 6, 8 (Trout Cr.); pl. 3, figs. 5, 7 (Succor Cr.), (not fig. 9, which is *M. reticulata*), 1936.

Odostemon simplex (Newberry) Cockerell. Berry, U.S. Geol. Surv. Prof. Paper 185-E, p. 112, pl. 23, fig. 1 only (not fig. 2, which is *Quercus hannibali* Dorf), 1934 (Hog Cr.).

Mahonia reticulata (MacGinitie) Brown. Axelrod, Carnegie Inst. Wash. Pub. 553, p. 255, pl. 43, fig. 7, 1944 (Alvord Cr.).

Axelrod, Carnegie Inst. Wash. Pub. 590, p. 60, pl. 3, fig. 1, 1950 (Napa).

Axelrod, Univ. Calif. Pub. Geol. Sci., vol. 33, pl. 8, fig. 16 (Aldrich Sta.); pl. 21, figs. 1-3 (Fallon); pl. 29, fig. 5 (Middlegate), 1956.

Axelrod, Univ. Calif. Pub. Geol. Sci., vol. 39, p. 234, pl. 48, fig. 2, 1962 (Chalk Hills).

Axelrod, Univ. Calif. Pub. Geol. Sci., vol. 51, p. 121, pl. 13, fig. 1, 1964 (Trapper Cr.).

Description: Leaflets elliptic to ovate, 2.5-7.0 cm long, 1.8-2.5 cm broad; apex acute, base asymmetrical, acute to blunt; petiolule 1.5 mm long, thick; midrib firm, thinning above, somewhat arched; 4-5 alternate secondaries and strong intersecondaries, both markedly camptodromous; tertiaries form irregular polygons within the looping secondaries and intersecondaries; tertiaries branch from loops to supply marginal teeth; teeth sharp, separated by 3-5 shallow sinuses; texture firm.

Discussion: The above-listed leaflets are all basically ovate to elliptic in outline, with a subequal base; the margin is shallowly dentate with sharp spines, the number variable depending on position of the leaflet on the rachis, and the venation is clearly pinnate.

The monograph on Mahonia (Ahrendt, 1961) shows that many living species are distinguished by characters that are not preserved as fossil--leaflets rigid or not; grey or green below; thick or thin; in 1-2 pairs or 3-5 pairs; broad or narrower. Other species are based (at least in part) on the number of marginal teeth, yet examination of herbarium specimens shows that in some species this is a variable character. In the case of the fossils here referred to M. macginitiei, there is considerable variation in number of teeth and depth of the sinuses. It seems unwise to subdivide these into 3 or 4 species on the basis of these features. The variation of the fossil material is matched by leaflets of the living M. aquifolium-sonnei-piperiana complex, species that otherwise are separated on the basis of features (leaflets greenish-grey or whitish-grey below; epapillose or papillose below) that are not preserved as fossil. Since leaflets of M. aquifolium (Pursh) Nuttall display considerable variation in number and size of marginal teeth and depth of sinuses, it sems best to include all this material in one polymorphic species, a taxon probably ancestral to these three closely-related living taxa.

Most investigators have previously compared these fossil leaflets with those produced by the living *M. aquifolium*, a shrub that inhabits the moister, northwestern mountainous section of California and adjacent Oregon and ranges north into British Columbia and east into the Idaho panhandle.

As noted previously (Axelrod, 1950a, p. 60), the fragmentary type specimens of *M. (Odostemon) hollicki* (Dorf) Arnold represent two taxa. One (Dorf, 1930, pl. 10, fig. 7) incomplete specimen appears to be *Ilex sonomensis* Dorf. The other (ibid., pl. 10, fig. 8) fragment may be a *Mahonia* allied to *aquifolium*, but is too incomplete to stand as a type and is therefore rejected. The new name for the fossil leaflets of *Mahonia* allied to *M. aquifolium* is *M. macginitiei*, based on the well-preserved specimens in the Trout Creek flora (MacGinitie, 1933, pl. 7, figs. 1, 3, 5).

It is a pleasure to name this species for Dr. Harry D. MacGinitie in recognition of his many important contributions to Tertiary paleobotany.

Occurrence: Trout Creek, holotype no. 585, paratype nos. 586-87; Middlegate, hypotype nos. 4336, 6591-93; Eastgate, hypotype no. 6981.

Mahonia reticulata (MacGinitie) Brown
(Plate 11, figs. 5, 10; pl. 26, figs. 1-5; pl. 27, figs. 1-2)

Clematis reticulata MacGinitie, Carnegie Inst. Wash. Pub. 416, p. 45, pl. 6, fig. 4, 1933.

Mahonia reticulata (MacGinitie) Brown, U.S. Geol. Surv. Prof. Paper 186, p. 175 (name only, not pl. 52, fig. 4, which is *Ilex sinuata* Chaney and Axelrod), 1937.

Axelrod, Univ. Calif. Pub. Geol. Sci., vol. 51, p. 121, pl. 13, fig. 1 only, 1964.

Graham, Kent State Univ. Bull., Research Series 9, p. 70, pl. 6, fig. 6, 1965 (refigure of Arnold's *M. hollicki*, pl. 6, fig. 3; see below).

Wolfe, U.S. Geol. Surv. Prof. Paper 454-N, p. N23, pl. 9, figs. 9, 10, 1964.

Mahonia hollicki (Dorf) Arnold. Arnold, Univ. Mich. Mus. Paleo. Contrib., vol. 5, p. 61, pl. 3, figs. 5 (Strode Rch.=Succor Cr.), 9 (Trout Cr.), 1936.

This species was founded on an entire-margined leaflet in the Trout Creek flora, and similar leaflets have been recovered from the Succor Creek flora and a few miles to the east. The large suites in the Eastgate and Buffalo Canyon floras show that the leaflets of this species range from entire to those with one or two to several very fine, widely spaced spinescent teeth scattered along the margin.

Among living western species, the leaflets of *M. reticulata* show relationship to those of *M. insularis* (=*M. pinnata* ssp. *insularis* (Munz)), a 5-7 m semivine confined now to the Channel Islands off Santa Barbara. However, the chief difference is that the fossil leaflets are mostly entire, a feature not present in *M. insularis*, which is a member of sect. *Aquifoliateae*. The leaflets of several species of sect. *Paniculateae*, subsect. *Laxiracemosae*, notably *M. andrieuxii* (H.&A.) Fedde, *chochoco* (Schidl.) Fedde, and *longipes* (Standley) Standley of Mexico, collectively show closer relationship to the fossil leaflets. The fossil species *M. reticulata* may represent an extinct member of this alliance.

Occurrence: Middlegate, hypotype nos. 6594-95, homeotype no. 6596; Eastgate, hypotype nos. 6982-88, homeotype nos. 6989-94.

Mahonia simplex (Newberry) Arnold
(Plate 11, fig. 1; pl. 27, figs. 3-6, 9)

Berberis simplex Newberry, U.S. Geol. Surv. Monogr. 35, p. 97, pl. 56, fig. 2, 1889 (Bridge Creek).

Odostemon simplex (Newberry) Cockerell. Chaney, Carnegie Inst. Wash. Pub. 346, p. 116, figs. 7, 8 (Gray Ranch), figs. 9, 11 (Cove Cr.), 1927.

MacGinitie, Carnegie Inst. Wash. Pub. 416, p. 55, pl. 9, fig. 1 (specimen on left), 1933 (Trout Creek).

Mahonia simplex (Newberry) Arnold, Univ. Mich. Mus. Paleo. Contrib., vol. 5, p. 58, pl. 1, figs. 1-3, 6, 7 (Trout Creek); pl. 2, figs. 1, 2 (Carter Creek = Succor Cr. loc.), 1936.

Chaney and Axelrod, Carnegie Inst. Wash. Pub. 617, p. 176, pl. 33, figs. 2, 5 (Stinking Water), 3, 6 (Mascall), 1959.

Numerous specimens in the Eastgate flora represent basal as well as medial and terminal leaflets of this species. The material is referred to *simplex* because the 2-3 sinuses are as deep as those in the type Bridge Creek material, the teeth are sharp-pointed and awn-like, and the venation is palmate.

There is considerable variation in the degree of incision of the leaflets in taxa of this alliance, and the lobes often are pungent. More than one species may be represented in the specimens referred to *simplex* and the allied *marginata*. However, this may simply illustrate our difficulty in recognizing taxa in the usual case where only one or two specimens are recovered from a given site. In addition, it seems clear that some of the differences in depth of sinuses and size and number of marginal teeth reflect differences in the environment that affect leaflet (or leaf) size. Among the species previously referred to *simplex*, the Trout Creek record (Arnold, 1936, pl. 1, figs. 1-3) may represent a different species, because the leaflets are smaller and have shallower sinuses. But until similar leaflets are found in other floras, it seems best to retain them in *simplex*.

Members of this alliance are related to several east Asian species, including *M. japonica* (Thunberg) DC and *M. lomariifolia* Takeda.

Occurrence: Middlegate, hypotype no. 6597; Eastgate, hypotype nos. 6995-99, homeotype nos. 7000, 7136.

Mahonia trainii Arnold

Mahonia trainii Arnold, Univ. Mich. Mus. Paleo. Contrib., vol. 5, p. 62, pl. 3, figs. 4, 6, 8, 1936.

Smith, Mich. Acad. Sci., Arts, and Letters Papers, vol. 27, p. 114, pl. 3, fig. 2, 1939.

Mahonia marginata (Lesquereux) Arnold. Axelrod, Univ. Calif. Pub. Geol. Sci., vol. 33, p. 295, pl. 29, fig. 2 only, 1956.

The above-cited specimen from the Middlegate flora is the only one recovered at that site. It is allied to M. marginata, but differs from it in having shallower sinuses and more numerous teeth. The present specimen is well matched by leaflets of the living M. nervosa (Pursh) Nuttall (see sheets 205551 and M048526 in the UC Herbarium, Berkeley). This shrub inhabits the coastward slopes of the Pacific states from central California northward into British Columbia and eastward into the humid Idaho panhandle. It is a member of Mahonia subsect. Nervosae, which is otherwise of east Asian distribution (Ahrendt, 1961, p. 327).

Occurrence: Middlegate, holotype no 4337.

Family CERATOPHYLLACEAE

Ceratophyllum praedemersum Ashlee
(Plate 22, figs. 5, 10)

Ceratophyllum praedemersum Ashlee, Northwest Sci., vol. 6, pl. 1, no. 2, 1932.

Brown, U.S. Geol. Surv. Prof. Paper 186-J, p. 175, pl. 45, fig. 32, 1937.

Several specimens of a water plant in the Eastgate flora with slender, curving leaves evidently disposed in whorls appear to represent this widespread aquatic genus. Another record of it is in the Latah flora, and an allied species is in the Shangwang flora, China (Hu and Chaney, 1940).

One of the fossil slabs with Ceratophyllum also has a root-scar of Nymphaeites, consistent with deposition of the associated flora in a water-body.

Occurrence: Eastgate, hypotype nos. 6977-78, homeotype nos. 6979-80.

Family NYMPHAEACEAE

Nymphaeites nevadensis (Knowlton) Brown
(Plate 11, figs. 6-7; pl. 22, fig. 9;
pl. 28, figs. 1, 3-4; pl. 29, fig. 9)

Nymphaeites nevadensis (Knowlton) Brown, Wash. Acad. Sci. Jour., vol. 27, p. 509, fig. 10, 1937 (see synonymy).
Chaney and Axelrod, Carnegie Inst. Wash. Pub. 617, p. 175, pl. 32, figs. 8, 9; pl. 33, figs. 9, 10, 1959.

The record of N. nevadensis includes detached root-scars of rhizomes in the Middlegate flora, while fragmentary leaf impressions, root scars, and rhizomes with attached rootlets are in the Eastgate flora.

Among living waterlilies, the present fossils compare well with Nuphar on the basis of the deeply cordate leaf in the Eastgate sample. In the western United States, N. polysepalum Engelmann is widely distributed from central California northward to Alaska.

When more complete material becomes available it should be possible to separate the different species of water-lily that are now grouped into Nymphaeites.

Occurrence: Middlegate, hypotype nos. 6598-99; Eastgate, hypotype nos. 7005-08, 7140.

Family PLATANACEAE

Platanus dissecta Lesquereux

Platanus dissecta Lesquereux, Mem. Harvard Mus. Comp. Zool., vol. 6, no. 2, pl. 10, figs. 4, 5, 1878.
Condit, Carnegie Inst. Wash. Pub. 553, p. 80, pl. 15, figs. 1, 4, 1944.
Axelrod, Univ. Calif. Pub. Geol. Sci., vol. 33, p. 297 pl. 29, fig. 1, 1956.

The only record of P. dissecta in floras of the Middlegate basin is the single specimen figured earlier from the middle Miocene floras of the western Great Basin, being known from only two others, the Pyramid and Fingerrock. It is rare in floras younger than 13 m.y., having been recorded only in the Aldrich Station flora (Axelrod, 1956). By contrast, its abundance in several floras of the Sierra Nevada suggests that decreased precipitation east of the range restricted its occurrence there.

The Fingerrock specimens, referred to _P. bendirei_ Lesquereux, appear to represent a new species. A large suite of leaves shows that there is a complete range of variation from those with long, attenuated lobes with numerous closely spaced teeth to those that are entire, one of which was figured by Wolfe (1964, pl. 4, fig. 1).

Occurrence: Middlegate, hypotype no. 4338.

Platanus paucidentata Dorf

Platanus paucidentata Dorf, Carnegie Inst. Wash. Pub. 412, p. 94, pl. 10, figs. 4, 9; pl. 11, fig. 1; pl. 12, fig. 1, 1930.

Axelrod, Carnegie Inst. Wash. Pub. 476, p. 174, pl. 5, figs. 4, 5, 1937.

Axelrod, Univ. Calif. Pub. Geol. Sci., vol. 33, p. 298, pl. 29, fig. 6, 1956.

No additional specimens of this species have been recovered at the Middlegate site, and it is not now known from the Eastgate locality. The relatively small leaf with entire margins is quite similar to leaves produced by _P. racemosa_ Nuttall in interior southern California.

Occurrence: Middlegate, hypotype no. 4339.

Family HYDRANGEACEAE

Hydrangea ovatifolius Axelrod, new species
(Plate 11, figs. 3, 8)

Celastrus sp. Dorf, Carnegie Inst. Wash. Pub. 476, p. 120, pl. 3, fig. 7, 1936.

Hydrangea bendirei (Ward) Knowlton. Chaney and Axelrod, Carnegie Inst. Wash. Pub. 617, p. 180, pl. 36, fig. 9 only, 1959.

Description: Leaflets ovate, 6.0-8.0 cm long, 4.0-4.5 cm broad; petiole over 1 cm long; apex acute, base rounded to blunt; midrib firm, stout; 6-8 alternate secondaries, looping evenly upward at medium angles, camptodrome; occasional strong intersecondaries diverging from secondaries to supply marginal area; tertiaries thin, irregular, enclosing an irregular mesh; margin with small serrate teeth; texture medium.

Discussion: The specimens of *Celastrus* and *Hydrangea* from the Hog Creek and Mascall floras are ovate in outline like the Middlegate leaves. Thus, they differ from the lanceolate leaves of *H. bendirei* (Ward) Knowlton and *H. knowltoni* (Berry) Chaney and Axelrod; the latter has consistently larger leaves and samaras than the Mascall species. The leaves of *H. ovatifolius* are rather similar to those of *H. reticulata* MacGinitie from the Weaverville flora, though that species is larger and has fewer secondaries. Another allied species is *H. miobretschneideri* Hu and Chaney (1940) from the Shanwang flora, northeastern China, but it also is larger than the present specimens and is more similar to *H. reticulata*. Two figured leaves of *H. bendirei* from the Trapper Creek flora (Axelrod, 1964) are essentially intermediate between the present species and *H. bendirei*.

The specimens of *H. ovatifolius* are generally similar to the ovate leaves commonly produced by *H. paniculata* Siebold and *H. heteromalla* D. Don of China and Japan.

Occurrence: Middlegate, holotype no. 6600, paratype no. 6601; Mascall, hypotype no. 3141; Hog Creek, hypotype no. 1220.

Family GROSSULARIACEAE

Ribes stanfordianum Dorf
(Plate 29, figs. 1, 3)

Ribes stanfordianum Dorf, Carnegie Inst. Wash. Pub. 412, p. 97, pl. 10, fig. 6, 1930.
Axelrod, Univ. Calif. Pub. Geol. Sci., vol. 39, p. 234, pl. 48, figs. 5, 6, 1962.

Three specimens in the Eastgate flora are broadly 3-lobed and have all the characters typical of currant. Among the numerous living species of the genus, R. sanguineum Pursh, R. nevadense Kellogg, and R. viscosissimum Pursh have leaves rather similar to the fossils. In view of the limited fossil material, it is not feasible to designate one living species as the closest analogue of the fossils, though the more rounded lobes of the latter two species imply a closer relationship to them.

Occurrence: Eastgate, hypotype nos. 7009-10.

Family ROSACEAE

Amelanchier grayi Chaney
(Plate 28, fig. 2; pl. 30, figs. 2, 3)

Amelanchier grayi Chaney, Carnegie Inst. Wash. Pub. 346, p. 120, pl. 14, figs. 3-5, 1927.

The Eastgate leaves are more like A. grayi Chaney than any other described species and are therefore referred to it. The leaves are broadly ovate to rotund, 3.5-4.0 (or more) cm long; slender petiole 1.0-1.3 cm preserved; apex rounded, base rounded; about 7 alternate to subopposite secondaries diverging at moderate angles, looping upward just inside margin in lower half of blade, then either dividing to supply marginal teeth with a tertiary or entering teeth directly;

tertiary venation in lower third cross-percurrent, becoming irregularly polygonal above, the polygons of quadriles enclosing open-ended veinlets; upper half of blade with sharply acute teeth; texture relatively thin.

These leaves differ from those of *A. alvordensis* Axelrod in having a more rotund shape, and they are larger. *A. scudderi* Cockerell from Florissant has fewer secondaries, they are not so looping, the venation does not appear as modern as in the present material, and the teeth of *scudderi* are more rounded. Leaves of *A. couleeana* (Berry) Brown from Oregon and Washington are regularly much larger than the present material, but otherwise seem similar in shape and basic characters.

The present fossils are well matched by leaves of the living *A. florida* Lindley of the northwest coast of California, extending to Alaska. In addition, leaves from the Rocky Mountains, commonly labeled *A. alnifolia* Nuttall, are also similar to the fossils.

Occurrence: Eastgate, hypotype nos. 7011-13, homeotype nos. 7014-15.

Cercocarpus antiquus Lesquereux
(Plate 13, figs. 10-12)

Cercocarpus antiquus Lesquereux, Mem. Harvard Mus. Comp. Zool., vol. 6, no. 2, p. 37, pl. 10, figs. 6-11, 1878.

Condit, Carnegie Inst. Wash. Pub. 553, p. 82, pl. 16, fig. 3, 1944.

Axelrod, ibid., p. 256, pl. 43, figs. 3, 4, 1944.

Axelrod, Univ. Calif. Pub. Geol. Sci., vol. 33, p. 299, pl. 28, figs. 12-14, 1956.

Leaves of this species have a rhombic outline, 7-10 essentially parallel secondaries, and relatively large size, usually more than 3.5 cm long.

This is clearly a member of the *C. betuloides* Nuttall complex, but is not closely allied to any one living species. Its relationships to other species are considered below, under the discussion of *C. ovatifolius*.

Occurrence: Middlegate, hypotype nos. 6602-04, homeotype nos. 6605-10.

Cercocarpus eastgatensis Axelrod, new species
(Plate 12, figs. 5-6; pl. 30, figs. 7-8;
pl. 31, figs. 1-14; pl. 32, figs. 1-4)

Cercocarpus holmesii (Lesquereux) Axelrod, Univ. Calif. Pub. Geol. Sci., vol. 33, p. 299, pl. 28, figs. 4, 9, 11, 1956.

This species is based on numerous leaves in the Eastgate flora, and a few specimens in the Middlegate flora are also referred to it.

Description: Leaves oblanceolate, narrowly obovate to narrowly elliptic; 1.8-5.5 cm long, average 3.0-3.5, and 0.6-1.5 cm broad, average about 1.0 cm in upper, widest part of blade; tips acute to rounded; base cuneate; petiole thick and short, 0.3-0.8 cm long, average about 0.5 cm; midrib straight and firm; 7-8 secondaries diverging at 10-25°, thinning only at margin or as enter teeth; tertiaries form very irregular cross-ties, quaternary venation very dense, polygonal, enclosing a dense mesh of veins with endings that have thick cross-ties; 3-5 teeth in distal part of blade, or occasionally nearly entire, and some with teeth reaching down to middle of blade; teeth broadly triangular and apiculate; texture firm and heavy.

Discussion: The leaves of *C. eastgatensis* range from quite small lanceolate specimens, similar to those previously referred to *C. holmesii* (Lesquereux) Axelrod, to larger obovate specimens more typical of the present suite. Inasmuch as the types of *C. holmesii* are known only from a few fragmentary specimens which do not display the variation of the present suite, it is not certain that that species is identical to the present one. Inasmuch as the locality has not been relocated and the variation of holmesii is not now determinable, it seems best to retain it as a separate taxon, one characterized by linear to elliptic leaves. The illustrated specimens (Lesquereux, 1887) seem inseparable from leaves in the Creede flora. Specimens of C. creedensis Axelrod from the Creede flora and C. myricaefolius (Lesq.) MacGinitie from Florissant are related to C. eastgatensis, but differ in their consistently greater length, and they are lanceolate-elliptic, not prominently obovate. All of these taxa, together with C. bea-anne Becker from the Ruby and Beaverhead basins, southwestern Montana (Becker, 1961, 1969), represent an evolutionary line that shows reduction in leaf size in a group of species that form a distinct series of the genus.

The question arises as to whether eastgatensis represents small leaves of C. ovatifolius. This is possible because C. blancheae, which is allied to ovatifolius, produces occasional axillary leaves somewhat like the fossils referred to C. eastgatensis, though they regularly are not so long as the fossils. The fact that the leaves referred to eastgatensis are abundant also implies that it probably is a different species. It can be argued that the fossil ovatifolius produced leaves similar to two living groups of species, and that in time narrower leaves like eastgatensis were produced in fewer number in the derived taxon blancheae. However, that is unlikely, to judge from the Cercocarpus leaves in the Golddyke flora 130 km (80 mi.) south. Fully 200 leaves of C. ovatifolius have been recovered there, yet only 5 small oblanceolate leaves similar to those of C. eastgatensis have been collected. On the basis of that record, the Golddyke species seems comparable to the living C. blancheae, whereas at Eastgate there appear to be two species, ovatifolius and eastgatensis, both of which are represented by relatively abundant leaves.

C. eastgatensis is allied to C. breviflorus Gray, C. parviflorus Nuttall, and C. paucidentatus Watson. These shrubs occur chiefly in the oak-piñon-juniper woodland, and enter the lower margins of forest in open areas. They frequently form shrub communities on steep, drier, south- and west-facing slopes.

Collection: Eastgate, holotype no. 7016, paratype nos. 7017-32, 7123-24; hypotypes 7125-36; Middlegate, hypotype nos. 4353-55, 6611-12.

Cercocarpus ovatifolius Axelrod, new species
(Plate 32, figs. 5-13)

Description: Leaves typically ovate to obovate; averaging 4.5-5.5 cm long and 2.5 cm broad; tip rounded to acute, base cuneate to acute; petiole 1.0-1.5 cm long, heavy; 6-8 essentially parallel secondaries, craspedodrome; margin with triangular, chiefly concave, apiculate teeth, a few with convex teeth; tertiaries irregularly subpercurrent, enclosing a fine, dense, irregular, polygonal mesh of quaternaries, the finer nervilles open-ended in the polygons; texture firm or heavy.

Discussion: Leaves of C. ovatifolius differ from those of C. antiquus Lesquereux in their ovate to obovate outline, as compared with the typically rhombic-elliptic shape in antiquus. Condit (1944b) compared antiquus leaves with those of C. blancheae (=alnifolius), but the latter are obovate to ovate as compared with leaves from the type locality at Table Mountain (also see Lesquereux, 1878, pl. 10, figs. 6-11).

C. ovatifolius leaves compare favorably with leaves of C. blancheae, though the latter average larger for the most part. A few leaves in the suite have teeth that are convex-acuminate rather than concave-acuminate. They suggest relationship with C. macrourus of the Klamath Mountain region, where it inhabits the lower, open borders of mixed-conifer forest. This raises the possibility that C. ovatifolius is ancestral to both blancheae and macrourus, a relationship supported by the occurrence of similar variation in the suites of leaves in the Fingerrock and Golddyke floras.

Some authors have compared _C. antiquus_ with the living _C. betuloides_, but the latter is a smaller-leaved species with fewer (3-6 as compared with 7-10) secondaries. Leaves of _C. cuneatus_ Dorf from the Mount Eden (5 m.y.), Napa (3.5 m.y.), and Santa Clara (2-3 m.y.) floras are much like _betuloides_. That _cuneatus_ originated earlier in the southwestern interior is indicated by its abundance in the Mint Canyon flora (12 m.y.), which was then situated some 250 km southeast, near the present site of Salton Sea. Present evidence thus indicates that _C. antiquus_ is an extinct species in the _betuloides_ line of evolution, characterized by its rhombic-elliptic shape, 7-10 secondaries, and relatively large size.

The question arises as to why the large leaves of _Cercocarpus_ in the Middlegate and Eastgate floras represent different species, with _antiquus_ at Middlegate and _ovatifolius_ at Eastgate. A similar relationship is indicated by fossil floras to the south. The suites of leaves in the Fingerrock and Golddyke floras are typically ovate to obovate, yet the younger Stewart Spring flora has the typical rhombic leaves of _antiquus_. The difference may reflect local environments that characterized these areas. The large _ovatifolius_ leaves in the Eastgate, Fingerrock, and Golddyke floras evidently represent a species adapted to a more mesic environment, as compared with that under which _C. antiquus_ lived in the Middlegate, Stewart Spring, and Table Mountain areas.

Collection: Eastgate, holotype no. 7033, paratype nos. 7034-42, homeotype nos. 7043-53.

Crataegus middlegatensis Axelrod
(Plate 12, figs. 2-3)

Crataegus middlegatei Axelrod, Univ. Calif. Pub. Geol. Sci., vol. 33, p. 300, pl. 29, figs. 3, 4, 1956.

The new collections at Middlegate have yielded a battered leaf and a spine to add to the two specimens previously representing this species. They are similar to those produced by living species, especially *C. chrysocarpa* Ashe and also *C. columbiana* Howell of the Pacific Northwest. Leaves of the present Rocky Mountain species *C. erythropoda* Ashe and *C. coloradensis* A. Nelson are also similar to the fossils.

Occurrence: Middlegate, holotype no. 4358, paratype no. 4359, hypotype nos. 6613-14.

Crataegus newberryi Cockerell
(Plate 30, fig. 5)

Crataegus flavescens Newberry, U.S. Geol. Surv. Mono. 35, p. 112, pl. 48, fig. 1, 1898.

Crataegus newberryi Cockerell, Amer. Mus. Nat. Hist. Bull., vol. 24, p. 95, 1908.

Chaney, Carnegie Inst. Wash. Pub. 346, p. 121, pl. 14, figs. 6, 10, 1927.

The mid portion of a deeply lobed leaf blade seems similar to *C. newberryi* from Bridge Creek and also is within the range of variation of *C. copeana* (Lesquereux) MacGinitie (1953) from Florissant. Both species have been compared with leaves of the living *C. pinnatifida* Bunge of eastern Asia, and the present specimen also is within its range of variation.

Occurrence: Eastgate, hypotype no. 7054.

Crataegus pacifica Chaney

Crataegus pacifica Chaney, Contrib. Walker Mus., vol. 2, p. 176, pl. 17, fig. 1, 1920.

Condit, Carnegie Inst. Wash. Pub. 553, p. 49, pl. 11, figs. 1-3, 1944.

Axelrod, Univ. Calif. Pub. Geol. Sci., vol. 33, p. 301, pl. 30, fig. 1, 1956.

A single, battered leaf previously reported from the Middlegate flora is the sole record of the species at this locality. As noted previously, it shows relationship to leaves of *C. cuneatus* Siebold and Zuccarini of central China and northern Japan. *C. monogyna* Jacquin and *C. oxycantha*

Linnaeus of eastern Europe and the Middle East also have similar leaves.

Occurrence: Middlegate, hypotype no. 4351.

Heteromeles sonomensis (Axelrod) Axelrod, new comb.
(Plate 12, figs. 7-8, 10;
pl. 30, figs. 1, 4, 6)

Photinia sonomensis Axelrod, Carnegie Inst. Wash. Pub. 553, p. 139, 1944 (see synonymy and discussion).
Axelrod, ibid., p. 258, pl. 45, fig. 1, 1944.

Several leaves in the Middlegate and Eastgate floras represent Heteromeles, sometimes referred to as Photinia in older California manuals; as presently recognized, the latter is an Asian genus. Leaves of Heteromeles have more widely spaced teeth, not closely spaced ones; the teeth are not often sickle-shaped as in Photinia; the teeth of Heteromeles are often directed outward or even basally; nervation of Photinia is denser and more complex than that of Heteromeles.

The present specimens are long-elliptic in outline, 7.0-8.0 cm long and 3.0-3.5 cm broad; tip rounded or blunt, base acute; midrib stout and tapering above; petiole thick and about 1 cm long; numerous secondaries diverging at about 45°, looping within margin to join secondaries above; sending tertiaries to the marginal teeth; tertiary mesh of irregular polygons; the finer quaternary mesh somewhat elongated, roughly parallel to secondaries, enclosing a finer mesh of veins with open cross-ties; margin with serrate teeth, some apparently blunt and possibly glandular, often directed outward, or occasionally basally; texture heavy.

The present specimens are similar to the larger leaves produced by H. arbutifolia Roemer, a common shrub in the woodland belt of California that contributes also to chaparral as well as to the lower margin of forest in open areas.

Occurrence: Middlegate, hypotype nos. 6622-24; Eastgate, hypotype nos. 7063-65; homeotype no. 7066.

Lyonothamnus parvifolius (Axelrod) Wolfe
(Plate 12, figs. 1, 4, 9; pl. 27, fig. 8;
pl. 29, fig. 8)

Lyonothamnus parvifolius (Axelrod) Wolfe. Axelrod, Univ. Calif. Pub. Geol. Sci., vol. 121, p. 207, 1980.
Comptonia parvifolia Axelrod, Univ. Calif. Pub. Geol. Sci., vol. 33, p. 287, pl. 8, figs. 13-15, pl. 28, fig. 10, 1956.

Additional collections of more complete specimens show quite clearly that these fossils represent Lyonothamnus, not Comptonia. The specimens from the Miocene Aldrich Station, Eastgate, Purple Mountain, and Buffalo Canyon floras and from the younger (Hemphillian) Truckee flora near Hazen, all differ from the Stewart Spring fossils (Wolfe, 1964) in having much smaller leaves, and the lobations on the leaflets are about half to two-thirds as broad. There is a probability that the small-leaved L. parvifolius may have been a shrub rather than a tree.

L. cedrusensis Axelrod (1981, p. 207), presently known only from the Stewart Spring flora, probably was chiefly a streamside tree in the sclerophyll woodland zone that dominated near the shore. L. cedrusensis is more nearly allied to the living L. asplenifolius Greene of the Channel Islands than is L. parvifolius, though L. cedrusensis differs from asplenifolius in having slenderer leaf lobes. The development of broader and thicker (?) leaf lobes of asplenifolius may reflect adaptation to increasing summer drought following the late Miocene.

Occurrence: Middlegate, hypotype nos. 4289, 6615-17, homeotype nos. 6618-21; Eastgate, hypotype nos. 7055-56, homeotype nos. 7057-62.

Prunus chaneyi Condit
(Plate 30, fig. 9)

Prunus chaneyi Condit, Carnegie Inst. Wash. Pub. 476, p. 263, pl. 5, figs. 4, 5, 1938.
Chaney and Axelrod, Carnegie Inst. Wash. Pub. 617, p. 185, pl. 36, fig. 4, 1959.
Axelrod, Univ. Calif. Pub. Geol. Sci., vol. 51, p. 124, pl. 14, figs. 5-8, 1964.

Several leaves in the Eastgate flora represent this species, which shows relationship to members of the P. demissa (Nuttall) Walpers-P. virginiana Linnaeus phylad. The leaves are broadly ovate with acuminate tips. The margins are finely serrate, secondaries loop within the margin and supply the marginal area with tertiaries that fork and diverge into the fine serrate teeth. One slender elliptic leaf in the Eastgate flora is also within the range of variation of the species.

The modern taxa allied to the fossil are forest species chiefly, descending along streambanks to lower levels and occurring frequently at seepages and springs.

Occurrence: Eastgate, hypotype no. 7068, homeotype nos. 7069-70.

Prunus moragensis Axelrod

Prunus moragensis Axelrod, Carnegie Inst. Wash. Pub. 553, p. 140, pl. 30, figs. 6, 8, 10, 1944.
Axelrod, Univ. Calif. Pub. Geol. Sci., vol. 33, p. 301, pl. 28, fig. 15, 1956.
Axelrod, Univ. Calif. Pub. Geol. Sci., vol. 34, p. 132, pl. 23, figs. 11-13, 1958.
Axelrod, Univ. Calif. Pub. Geol. Sci., vol. 39, p. 235, pl. 49, fig. 5, 1962.

A single, oblong-elliptic leaf of this species was collected initially at Middlegate, but no additional specimens have been recovered there. The leaf has numerous alternate secondaries that diverge at moderate angles and loop upward along the margin to divide into tertiaries that supply the numerous crenately-serrate teeth. Among modern species, _P. emarginata_ (Douglas) Walpers produces leaves similar to the fossil. This shrub or small tree is distributed from sea level to high montane sites, ranging from California northward into British Columbia and eastward into Idaho.

The specimen figured as _P. moragensis_ from Trout Creek (Chaney and Axelrod, 1959, pl. 36, fig. 5) is incorrectly illustrated: the veins are looping and not straight, and the margin is serrate. This specimen is a leaflet and is herewith transferred to _Rosa harneyana_ Chaney and Axelrod.

Occurrence: Middlegate, hypotype no. 4360.

Sorbus idahoensis Axelrod, new species
(Plate 29, figs. 2, 6-7, 10)

Sorbus cassiana Axelrod, Univ. Calif. Pub. Geol. Sci., vol. 51, p. 125, pl. 14, fig. 3 only (not pl. 14, fig. 4, which is _Alnus_), 1964.

Rhus alvordensis Axelrod, Univ. Calif. Pub. Geol. Sci, vol. 33, p. 305, pl. 30, figs. 7, 8, 1956.

Axelrod, Univ. Calif. Pub. Geol. Sci., vol. 51, p. 125, pl. 14, fig. 11, 1964.

Examination of the type specimens of _Sorbus_ from Trapper Creek shows that the specimen designated as the holotype (Axelrod, 1964, pl. 14, fig. 4) is a broken, mangled leaf of _Alnus_, not two partial leaflets of _Sorbus_. For this reason, a new type must be selected for this _Sorbus_ species.

Description: Leaflets regularly oblong to oblong-lanceolate, acute below and with apiculate tips; 4.8 cm long and 1.6 cm wide in lower part, gradually thinning above; midrib medium, often slightly curved; numerous secondaries diverging at moderate angles, gently looping to wavering, tertiaries forming an irregular polygonal mesh enclosing coarse 4th-order veinlets; the 5th-order veinlets very fine, free-ending, irregular; margin with single-serrate teeth, usually more strongly developed on one side; texture firm.

Discussion: The leaflet from Trapper Creek is inseparable from those in the Middlegate and Eastgate floras, and they are therefore referred to the same species. They differ from *S. harneyensis* Axelrod from Alvord Creek in having a more slender shape, and the margin has fewer teeth. A larger suite of *Sorbus* leaflets and nearly complete leaves are in the nearby Buffalo Canyon flora and will be illustrated with that report.

The material in hand compares favorably with leaves and leaflets produced by *S. acuparia* Linnaeus of Eurasia and *S. pohuashanensis* (Hance) Hedley of northern China.

Occurrence: Trapper Creek, syntype no. 8533; Middlegate, homeotype nos. 6625-26; Eastgate, hypotype nos. 7071-74, homeotype nos. 7075-80.

Family LEGUMINOSAE

Gymnocladus dayana (Knowlton) Chaney and Axelrod
(Plate 15, figs. 2-3)

Gymnocladus dayana Chaney and Axelrod, Carnegie Inst. Wash. Pub. 617, p. 187, pl. 37, figs. 7-9, 1959 (see synonymy).

A single leaflet and its counterpart in the Middlegate flora appear referable to this species. The leaflet is broadly ovate, with an acute tip and a rounded base; the midrib is firm, thinning above, the very base missing. The secondaries diverge at moderate (45°) angles, and loop gently upward along the margin, sending off tertiaries to the marginal area. The tertiaries within the blade proper are irregularly polygonal and enclose a similar mesh of quaternaries. The present specimen resembles leaflets produced by *G. dioica* K. Koch of the central United States.

Occurrence: Middlegate, hypotype nos. 6627a-b.

Robinia californica Axelrod
(Plate 13, figs. 1-4; pl. 33, figs. 1-2)

Robinia californica Axelrod, Carnegie Inst. Wash. Pub. 516, p. 114, pl. 9, figs. 10-12, 1939.
Condit, Carnegie Inst. Wash. Pub. 553, p. 84, pl. 17, figs. 4, 5, 1944.
Axelrod, Univ. Calif. Pub. Geol. Sci., vol. 33, p. 304, pl. 14, fig. 14; pl. 30, figs. 2-6, 1956.

The new collections at Middlegate have added four legume pods and two leaflets to the earlier record of the species at this site. In addition, one leaflet of this species is in the Eastgate flora.

The leaflets and pods are similar to those of R. neomexicana Gray. R. pseudoacacia Linnaeus has larger leaflets and pods than the present fossils. R. neomexicana is a common shrub or small tree at the forest margin in the southwestern United States and adjacent Mexico, ranging well down into the oak-juniper-piñon belt.

Occurrence: Middlegate, hypotype nos. 6628-31, homeotype nos. 6632-33; Eastgate, hypotype nos. 7067, 7081.

Family ACERACEAE

Acer middlegatensis Axelrod
(Plate 14, figs. 10-12)

Acer middlegatei Axelrod, Univ. Calif. Pub. Geol. Sci., vol. 33, p. 307, pl. 31, figs. 7-11, 1956.

A number of additional specimens of this species have been recovered from the Middlegate site. They show that there is no significant overlap in size of leaves or samaras with those produced by A. osmonti Knowlton and A. gigas Knowlton, respectively (Knowlton, 1902). It thus seems best to retain the Middlegate species distinct from the Oregon taxon, even though both are similar to leaves and samaras produced by the eastern silver maple, A. saccharinum Linnaeus.

Although *A. osmonti* Knowlton, which includes *A. gigas* Knowlton, has been recorded at a number of sites in Oregon, Washington, and Idaho (see Chaney and Axelrod, 1959, p. 191, under *A. bendirei*), it is now known only from the Pyramid flora in the Great Basin area. Significantly, its leaves and samaras are much larger than those of *A. middlegatensis*, being the size of the type specimens from the Mascall flora.

Occurrence: Middlegate, holotype no. 4388, paratype nos. 4389, 4392-95, hypotype nos. 6643-45, homeotype nos. 4390-91, 6646-48.

Acer negundoides MacGinitie
(Plate 14, figs. 7-9)

Acer negundoides MacGinitie, Carnegie Inst. Wash. Pub. 416, p. 62, pl. 11, figs. 2-3, 1935.

Chaney and Elias, Carnegie Inst. Wash. Pub. 476, p. 42, figs. 10-11, 1936.

Brown, Jour. Paleontol., vol. 9, p. 580, pl. 69, figs. 9-11, 1935.

Chaney, Carnegie Inst. Wash. Pub. 553, p. 320, pl. 53, figs. 2, 4, 1944.

Acer minor Knowlton. Axelrod, Univ. Calif. Pub. Geol. Sci., vol. 33, p. 308, pl. 32, figs. 5-7, 1956.

Chaney and Axelrod, Carnegie Inst. Wash. Pub. 617, p. 194, pl. 41, figs. 3-6, 1959.

J. A. Wolfe is now revising the fossil maples, and has decided that the samara representing *A. minor* Knowlton is too incomplete to stand as a type. The next available name for fossils of the *negundo* alliance is *A. negundoides* MacGinitie, and is therefore used here.

A number of additional samaras have been recovered from the Middlegate site, together with two battered leaflets. This material, like that cited above, resembles leaflets and samaras of *A. negundo* Linnaeus of the United States and also those of the allied *A. henryi* Pax of Hupeh and Szechuan provinces, central China.

The occurrence of _A_. _negundoides_ in some floras and its absence from others is not readily accounted for. In the western United States it occurs chiefly along stream borders and river bottoms in relatively warmer areas. This may explain its absence from the nearby Eastgate flora, in a cooler site than Middlegate, but scarcely accounts for its abundance in the Buffalo Canyon flora 20 km southeast, where the climate was cooler than at Eastgate. In Oregon it is absent from the Miocene Blue Mountains flora, which represents a conifer-deciduous hardwood forest, yet it occurs in the Thorn Creek flora, which represents generally similar vegetation.

Occurrence: Middlegate, hypotype nos. 4398-4400, 6637-39, homeotype nos. 4396-97, 4401-03, 6640-42.

Acer _nevadensis_ Axelrod, new species
(Plate 13, figs. 5-7; pl. 33, figs. 4-5)

A single leaf and two samaras comprise the record of this maple in the Middlegate flora, and a samara and its counterpart are in the Eastgate.

Description: Leaf circular in outline, 3-lobed, the lobes diverging at 45°, arching at moderately higher angles in the far distal part of the lateral lobes; leaf blade 3.0 cm long and 3.5 cm broad at widest part; petiole missing; base broadly obtuse, tips acute; the primaries in each lobe slender, the laterals curving outward in distal portion; secondaries diverge at moderate angles, looping upward; tertiaries form a coarse polygonal mesh enclosing open-ended quadriles; margin with small serrate teeth in upper half of blade; texture medium.

Samaras 2.3-2.5 cm long, seed generally oval in outline, 7-8 mm in length and breadth; line of dehiscence at 45°; seed conspicuously grooved, leaving a deep furrow in matrix; wings broadly attached to seed, without a deep sinus below, widest in middle part, tip obtuse, wing heavily veined, with veins diverging downward distally to supply marginal area.

Discussion: The fossils of *A. nevadensis* are similar to leaves and samaras of *A. diffusum* Greene and also to *A. torreyi* Greene, both generally considered varieties of *A. glabrum* Torrey (F. J. Smiley, 1921, p. 261-262). *A. diffusum* is a widely distributed shrub in the mountains of the Great Basin, ranging to the east slope of the Sierra Nevada. *A. torreyi* replaces it in the higher parts of the Sierra Nevada and Peninsular Ranges. Both taxa range down into the piñon-juniper belt.

A. coloradense MacGinitie from Florissant (MacGinitie, 1953) is allied to *A. nevadensis*, but differs in being more deeply cleft. It compares well with *A. glabrum* var. *neomexicanum* (Greene) Kearney and Peebles of the southern Rocky Mountains, which tends to be trifoliate. Recalling that the Florissant maple is 34 m.y. old, and that the small-leaved maples in the Middlegate basin floras are 18.5 m.y. old, it seems appropriate to ask: Is it possible for a variety to persist for such a long time? This also applies to both the living *A. diffusum* and *A. torreyi*, which have been considered varieties of *A. glabrum*. A similar problem is posed by a number of other woody taxa that have fossil records. *Lyonothamnus asplenifolius* is considered a variety of *L. floribundus*, yet both have been in essentially modern form since 12-13 m.y. ago. *Rhamnus crocea*, as well as the var. *ilicifolia*, have fossil records in the late Miocene. Both are very distinct taxa today, with *R. ilicifolia* Kellogg often a large shrub or small tree with large leaves, whereas *R. crocea* is a semi-prostrate, divaricately branched shrub with very small leaves. The problem is compounded by the named varieties of *Cercocarpus betuloides*; var. *blancheae* appears to have been a distinct species for at least 16-18 m.y., and suites of smaller leaves like var. *glaber* occur in the late Miocene (12-13 m.y.).

As judged from the fossil record, I believe that these recognized varieties are, in fact, full-fledged species.

Occurrence: Middlegate, holotype no. 6649, paratype nos. 6650-51; Eastgate, paratype nos. 7091-91a (counterpart).

Acer oregonianum Knowlton
(Plate 13, figs. 8-9; pl. 15, figs. 4-5;
pl. 34, figs. 1-2)

Acer oregonianum Knowlton, U.S. Geol. Surv. Bull. 204, p. 75, pl. 13, figs. 5, 7, 8 (not fig. 6, which is A. scottiae MacGinitie), 1902.

Chaney and Axelrod, Carnegie Inst. Wash. Pub. 617, p. 195, pl. 41, figs. 11-14, 1959 (see synonmym).

Acer alvordensis Axelrod. Axelrod, Univ. Calif. Pub. Geol. Sci., vol. 33, p. 306, pl. 30, figs. 9-11, 1956.

Acer macrophyllum Wolfe, not Pursh, U.S. Geol. Surv. Prof. Paper 454-N, p. N29, pl. 5, figs. 4-6, 1964 (and all items in synonymy).

Numerous samaras and a few leaves have been recovered from the Middlegate site, two samaras and a few leaves are in the Eastgate deposit. The additional leaf specimens collected at Middlegate show that they are not so deeply lobed as the specimen figured previously as A. alvordensis, so that record is reduced to synonymy under A. oregonianum.

The samaras are among the most readily recognized of all maples, in view of their large size and the characteristic constriction of the wing under the seed. The samaras and leaves are similar to those produced by the living A. macrophyllum Pursh. This is a frequent tree in moist sites in the upper part of the evergreen sclerophyll forest and in the lower part of the mixed-conifer forest. Its site in a warmer area probably favored the greater abundance of A. oregonianum near Middlegate, as compared with the Eastgate flora.

Occurrence: Middlegate, hypotype nos. 4369, 4373-74, 6652-55, homeotype nos. 4370-72, 4375-79, 6656-63; Eastgate, hypotype nos. 7085-86, homeotype nos. 7087-89.

Acer scottiae MacGinitie
(Plate 15, figs. 10-11)

Acer scottiae MacGinitie, Carnegie Inst. Wash. Pub. 416, p. 62, pl. 12, fig. 4; pl. 11, figs. 4, 8, 1933. Chaney and Axelrod, Carnegie Inst. Wash. Pub. 617, p. 196, pl. 41, figs. 6-10; pl. 42, fig. 1, 1959 (see synonymy).

This distinct species has leaves with spreading, entire-margined lobes and a samara with a large seed that is sharply truncated normal to the axis of the wing. It is a member of the *pictum* group, showing relationship to *A. pictum* Thunberg as well as its close allies, *A. cissifolium* C. Koch, *A. truncatum* Bunge, and others in China and Japan.

Whereas the species has been recorded at a number of sites in the Columbia Plateau and bordering region, this is the first record of it in the central Great Basin region.

Occurrence: Middlegate, hypotype nos. 6664-65, homeotype no. 6666.

Acer tyrrelli Smiley
(Plate 14, figs. 1-6; pl. 33, figs. 3, 6-7, 9-12; pl. 34, figs. 3-4)

Acer tyrrelli Smiley, Univ. Calif. Pub. Geol. Sci., vol. 35, p. 227, pl. 13, figs. 3, 5, 1963.

Acer bolanderi Lesquereux. Axelrod, Univ. Calif. Pub. Geol. Sci., vol. 51, p. 127, pl. 16, figs. 10-12, 1964.

Acer arida Axelrod, Univ. Calif. Pub. Geol. Sci., vol. 33, p. 307, pl. 31, figs. 1-6, 1956.

The relatively abundant leaves of this species in the Eastgate flora are palmately 3- and 5-lobed. Size is variable from small leaves 3 cm long to larger ones 6 cm long, and all of them are deeply lobed. The base may range from cordate to truncate. The leaves are well matched by those of *A. grandidentatum* Nuttall, which are primarily 5-lobed as compared with the other allied species, notably *A. floridum* Pax, *A. leucoderme* Small, and *A. saccharum* Marshall, which have 3 primary lobes, with 2 smaller basal ones (when present) being subordinate. Several samaras in the Eastgate flora are similar to those of the species of the saccharum alliance, including grandidentatum, brachypterum, and others.

The Eastgate suite includes small leaves like the Middlegate specimens previously identified as A. arida Axelrod, and several additional specimens from Middlegate are larger than those previously collected. The rarity of larger leaves at Middlegate may well reflect a warmer, drier climate there, as well as the problem of distant transport and destruction prior to reaching the area of plant accumulation. The small Middlegate and Eastgate leaves are similar to those of the southern population of A. grandidentatum designated as A. brachypterum Wooton and Stanley.

In view of the occurrence of both 3- and 5-lobed leaves in the large Eastgate collection, one might be inclined to refer them to two species, A. bolanderi Lesquereux and A. tyrrelli Smiley. I believe that the fossil species was variable in leaf form and resembled two living taxa, grandidentatum proper (5-lobed) and its variety brachypterum (3-lobed). It is possible that A. tyrrelli was ancestral to the two modern taxa.

The small leaves of A. arida Axelrod described previously from the Tehachapi and Piru Gorge floras are retained as valid taxa, because they are most similar to the leaves of A. brachypterum, a species that persisted in southern California into the early Pleistocene (Axelrod, 1966b).

A. grandidentatum ranges through the central and southern Rocky Mountains, reaching to the front of the Wasatch Mountains and related ranges at the margin of the Great Basin. The tree descends along stream margins from the lower mixed-conifer forest well down into the conifer woodland belt dominated by Pinus edulis and Juniperus osteosperma. Among its associates are species of Acer (negundo, glabrum), Amelanchier, Betula, Cercocarpus, Populus, Prunus, Ribes, Rosa, and Salix with allied species in the fossil floras.

Occurrence: Middlegate, hypotype nos. 6667-72, homeotype nos. 6673-78; Eastgate, hypotype nos. 7082-84, 7090, 7092-96, homeotype nos. 7097-7105, 7083.

Family MELIACEAE

Cedrela trainii Arnold
(Plate 15, figs. 6-8)

Cedrela trainii Arnold, Amer. Midland Natur., vol. 17, p. 1018, figs. 1, 2, 1936.
Chaney and Axelrod, Carnegie Inst. Wash. Pub. 617, p. 189, pl. 38, figs. 3, 5-9, 1959 (see synonymy).

Three samaras in the Middlegate flora display the typical features of Cedrela. The slender seeds are attached to the wing almost tangentially across its upper part, the wing reaches under the seed where it is sharply truncated, and the outer edge of the wing shows a prominent marginal scar where pressure from adjoining seeds has left a conspicuous mark.

Cedrela ranges today from the tropics into the mild, temperate oak-pine country of central Mexico. Judging from the rarity of its samaras and the absence of its leaflets at Middlegate, Cedrela probably occupied sheltered, mesic sites well removed from the immediate area of plant accumulation.

Occurrence: Middlegate, hypotype nos. 6634-36.

Family HIPPOCASTANACEAE

Aesculus preglabra Condit
(Plate 34, figs. 5-7)

Aesculus preglabra Condit, Carnegie Inst. Wash. Pub. 553, p. 50, pl. 10, fig. 3; pl. 11, figs. 4, 6, 1944.

Several leaflets in the Eastgate flora resemble those produced by the living A. glabra Willdenow, and on this basis are referred to the fossil A. preglabra, which also resembles it. Two of the illustrated fossils of A. preglabra Condit (1944a, pl. 10, fig. 3, and pl. 11, fig. 6) from Remington Hill have the obovate outline of terminal leaflets which are not represented in the Eastgate suite except for one battered specimen. The lateral leaflet figured by Condit (1944a, pl. 11, fig. 4) is much like some of our specimens.

The Eastgate fossils differ from leaflets of A. californica (Spach.) Nuttall in marginal features, for the teeth are blunt to crenate in the California species, not sharply serrate as in the fossils and in the living A. glabra of the eastern United States.

Occurrence: Eastgate, hypotype nos. 7084, 7106-08, homeotype nos. 7109-10.

Family RHAMNACEAE

Ceanothus precuneatus Axelrod

Ceanothus precuneatus Axelrod, Carnegie Inst. Wash. Pub. 516, p. 120, pl. 11, fig. 7, 1939.

Axelrod, Carnegie Inst. Wash. Pub. 553, p. 261, pl. 45, figs. 2, 3, 1944.

Axelrod, Univ. Calif. Pub. Geol. Sci., vol. 33, p. 308, pl. 32, fig. 8, 1956.

A single leaf remains the sole record of this species in the Middlegate flora. In view of the preference of its living ally, C. cuneatus Nuttall, for dry slopes, its rarity in the record is not surprising.

Occurrence: Middlegate, hypotype no. 4404.

Rhamnus precalifornica Axelrod

(Plate 33, fig. 8)

Rhamnus precalifornica Axelrod, Carnegie Inst. Wash. Pub. 516, p. 122, pl. 11, figs. 1, 2, 1939.

Axelrod, Carnegie Inst. Wash. Pub. 553, p. 144, 1944.

Axelrod, Carnegie Inst. Wash. Pub. 590, p. 155, pl. 3, figs. 14, 15, 1950.

A single specimen in the Eastgate flora seems to represent this species. The leaf impression is incomplete, with the upper fourth of the blade missing. However, the typically elliptic outline, the relatively numerous parallel camptodrome secondaries, the strong cross-percurrent venation tertiary, and the thick petiole are readily matched by the leaves of _R_. _californica_ Eschscholtz, a highly variable species over its range. The slight asymmetry of the leaf is not uncommon in the modern species, and the margin may range from finely serrate to entire; the fossil leaf is entire.

_R_. _california_ is widely distributed, ranging from the coastal forests into the sclerophyll woodland, and up into the lower half of the mixed-conifer forest in the Sierra Nevada.

_R_. _columbiana_ Chaney and Axelrod (1959, p. 196) is an allied, closely similar species. When additional material becomes available, it may be desirable to merge them.

Occurrence: Eastgate, hypotype no. 7112.

Family MYRTACEAE

Eugenia _nevadensis_ Axelrod, new species
(Plate 34, fig. 9)

Description: Leaves long-lanceolate, curved; 7 cm long and 1.5 cm broad in widest basal part; tip acute and base rounded; midrib thick, tapering above; petiole 1 cm long and heavy; 16+ secondaries, alternate to subopposite, diverging at 45° (above) to 70° (below), looping upward to form a continuous marginal vein the length of the blade; inter-secondaries thin, irregular in outward course, some looping up, others down; tertiaries thin, forming an irregular polygonal mesh; finer veins not visible, probably because of thickness of leaves; margin entire, texture firm or thick.

Discussion: Three leaves in the Buffalo Canyon flora and a very slender one in the Eastgate flora are myrtaceous, as shown by the marginal vein. This is formed by the secondaries that branch just within the margin; the branch then joins the adjacent secondary to form the marginal vein. These specimens are referred to _Eugenia_ on the basis of their similarity to leaves in the Florissant (MacGinitie, 1953, pl. 71) and Green River floras (MacGinitie, 1969, pl. 18 and 19). Those species have larger leaves and no doubt represent different species, but otherwise are quite similar in basic features. MacGinitie noted their resemblance to leaves of _E. fluvatilis_ Hemsley and _E. jambos_ Linnaeus.

Apart from its occurrence in the present floras, _E. nevadensis_ also occurs in the Fingerrock flora to the south and in the Carson Pass flora on the Sierran crest. _Eugenia_ was clearly a relict in these Miocene floras.

Occurrence: Eastgate, holotype no. 7113.

Family STYRACACEAE

Styrax middlegatensis Axelrod

Styrax middlegatei Axelrod, Univ. Calif. Pub. Geol. Sci. vol. 33, p. 309, pl. 32, fig. 9, 1956.

This species is represented by the single leaf collected initially at Middlegate; no additional specimens have been recovered there. The allied living _S. californica_ Munz and Johnston is a frequent member of the chaparral vegetation of the Sierra Nevada and southern California, reaching up to the lower margin of forest. It prefers drier, sunnier slopes, and such a habitat preference may account for the rarity of its remains in the Middlegate record.

Occurrence: Middlegate, holotype no. 4405.

Family EBENACEAE

Diospyros oregonianum (Lesquereux) Chaney and Axelrod (Plate 15, fig. 9)

Diospyros oregoniana (Lesquereux) Chaney and Axelrod, Carnegie Inst. Wash. Pub. 617, p. 200, pl. 43, fig. 7, 1959.

The mid portion of a large ovate leaf in the Middlegate flora is identified as this species on the basis of its exceptionally well-preserved nervation. The strong, irregular tertiaries that diverge largely normal to the looping secondaries enclose a mesh of irregular polygonal quaternaries, and the enclosed 5th-order mesh includes numerous open veinlets. The venation is matched closely by leaves of *D. virginiana* Linnaeus of the eastern United States and by *D. kaki* Linnaeus of China and Japan.

Its rarity in the Middlegate flora implies that it probably lived upstream and was battered during transport to the lake.

Occurrence: Middlegate, hypotype no. 6679.

Family ERICACEAE

Arbutus prexalapensis Axelrod
(Plate 13, fig. 13; pl. 15, fig. 1; pl. 34, fig. 10)

Arbutus prexalapensis Axelrod, Carnegie Inst. Wash. Pub. 516, p. 124, pl. 11, fig. 11, pl. 12, fig. 2, 1939.
Axelrod, Carnegie Inst. Wash. Pub. 590, p. 113, pl. 3, figs. 11, 14, 1950.
Axelrod, Univ. Calif. Pub. Geol. Sci., vol. 33, p. 310, pl. 32, figs. 1, 2, 4 (not fig. 3, which is *Salix owyheeana* Chaney and Axelrod)
Axelrod, Univ. Calif. Pub. Geol. Sci., vol. 59, p. 77, pl. 18, figs. 11, 12, 1966.

The new Middlegate collection includes several additional leaves that show relationship to members of the genus now in areas well to the south. Like the leaves previously collected, they are entire, with anastomosing secondary venation. In shape, secondary and finer nervation, and margin, the fossils resemble leaves of *A. arizonica* (A. Gray) Sargent, a common member of the upper evergreen sclerophyll woodland in the mountains of southern Arizona, southwestern New Mexico, and northern Mexico.

More numerous specimens are in the Eastgate flora. They average somewhat larger than those at Middlegate, and two or three are coarsely crenate-serrate, with widely spaced teeth. The larger leaves at Eastgate correspond with the inferred difference in local climate, living under moister conditions, as indicated by the more abundant representation of forest taxa than at Middlegate.

A. trainii MacGinitie is an allied, larger-leaved species common in middle Miocene floras to the north. Its leaves are similar to those of *A. xalapensis* H.B.K. in the mesic parts of central and southern Mexico. Several previously identified "species" of *Arbutus* in Mexico probably represent no more than variations of the polymorphic *xalapensis*. The smaller-leaved *A. arizonica* and *A. texana* appear to be taxa that were derived from an *A. xalapensis* plexus (i.e. *trainii*) by adaptation to a drier, cooler climate than that now to the south in Mexico, where *xalapensis* finds optimum development in the montane forests. That the smaller-leaved species had already evolved by the later Eocene is apparent from the Copper Basin record (Axelrod, 1966a). The southward retreat of *Arbutus* from Nevada represents a response to post-Miocene developments of colder winters, a drier climate, and the elimination of summer rainfall.

Occurrence: Middlegate, hypotype nos. 4406-09, 6680-81, homeotype nos. 4410-11, 6682-83; Eastgate, hypotype no. 7014, homeotype nos. 7015-20.

Family OLEACEAE

Fraxinus coulteri Dorf
(Plate 34, fig. 8)

Fraxinus coulteri Dorf, Carnegie Inst. Wash. Pub. 476, p. 123, pl. 3, figs. 3, 4, 1936.

Chaney and Axelrod, Carnegie Inst. Wash. Pub. 617, p. 200, pl. 44, fig. 5, 1959.

Axelrod, Univ. Calif. Pub. Geol. Sci., vol. 33, p. 311, pl. 32, fig. 12, 1956.

No additional specimens of this distinctive samara were recovered during the more recent excavations of Middlegate. In view of the habitat that the species no doubt occupied, this is surprising, for its closest living ally, *F. oregona*

Nuttall, is a common streambank tree in the lower part of mixed-conifer forest and in sclerophyll vegetation as well.

Occurrence: Middlegate, hypotype nos. 4413-14; Eastgate, hypotype no. 7121.

Fraxinus millsiana Axelrod

Fraxinus millsiana Axelrod, Univ. Calif. Pub. Geol. Sci., vol. 33, p. 311, pl. 32, figs. 10, 11, 1956.

No additional specimens of this unique species have been recovered from the Middlegate sediments. These samaras, characterized by relatively broad, oval wings that nearly clasp the entire seed, are similar to those of the living F. anomala Torrey, a rather large shrub found in the pinon-juniper woodland and chaparral communities in southeastern California, southern Utah, and Arizona.

Occurrence: Middlegate, holotype no. 4415, paratype no. 4416, homeotype no. 4417.

SYSTEMATIC REVISIONS

The preceding Systematic Descriptions include a number of changes in the disposition of species in previously described floras. This has resulted in part from new, larger collections that illustrate more clearly the variation of the fossil species and disclose more definitively the nature of the venation in some species described previously from poorly preserved material. To determine the affinity of the taxa, leaf variation in the fossil suites has been compared with that of the living species, as judged from herbarium material. This procedure is more reliable than one based on a single cleared leaf, for it cannot display the foliar variation of a species.

The following changes are grouped into two parts. Part I includes changes made on the preceding pages. Part II presents changes made in other floras that were consulted during the course of this study.

PART I

Thuja dimorpha (Oliver) Chaney & Axelrod. Axelrod, 1956, pl. 4, fig. 24; pl. 12, figs. 1-4; pl. 18, figs. 1, 2; pl. 25, figs. 2, 3 = Chamaecyparis cordillerae Edwards & Schorn 118

Torreya nancyana Axelrod, 1956, pl. 28, fig. 6 only = Abies scherri Axelrod 109

Tsuga heterophylla Wolfe (not Sargent), 1964, pl. 6, figs. 15, 16, 20, 21, 24 = Tsuga mertensioides Axelrod 117

PART II

Alnus corylina Knowlton and Cockerell

Alnus corylina Knowlton and Cockerell, U.S. Geol. Surv. Bull. 696, p. 63, 1919.

Wolfe, U.S. Geol. Surv. Prof. Paper 398-B, p. B19, pl. 6, figs. 2, 5; fig. 7, 1966 (all items in synonymy).

Alnus cappsii (Hollick) Wolfe, ibid., p. B19, pl. 6, figs. 1, 4; pl. 7, figs. 2, 6; fig. 6, 1966 (all items in synonymy).

Alnus barnesii Wolfe, ibid., p. B19, pl. 5, figs. 5, 7; fig. 8, 1966.

Alnus schmidtae Wolfe, ibid., p. B19, pl. 5, fig. 1; fig. 10, 1966 (all items in synonymy).

As discussed under the family Betulaceae, leaves of all these Alaskan Miocene alders display variation similar to leaves of the living A. tenuifolia-A. incana complex, and are therefore grouped under one species.

Alnus smithiana Axelrod (part). Axelrod, 1956, p. 288, pl. 7, figs. 5, 10-12 (not figs. 13-14, which are Alnus walkeri Axelrod) = Betula smithiana (Axelrod) Axelrod

The leaves and winged seeds from Aldrich Station represent a birch allied to the living B. fontinalis Sargent. The cones that were originally assigned to A. smithiana do represent alder.

Alnus smithiana Axelrod (part). Axelrod, 1956, p. 288, pl. 7, figs. 13-14 only = *Alnus walkeri* Axelrod

These cones are alder, but leaves certainly referable to *Alnus* have not yet been recognized in the Aldrich Station flora.

Betula thor Knowlton. Axelrod, 1962, p. 232, pl. 47, figs. 1-4 = *Betula smithiana* (Axelrod) Axelrod

These specimens from the Chalk Hills flora are similar to those in the Aldrich Station assemblage. They more nearly resemble the living *B. fontinalis* Sargent than *B. papyrifera* (see *Alnus smithiana* Axelrod, above). The leaves of *B. papyrifera* usually are not so lobed along the margin, and thus give it a more nearly regular outline.

Colubrina sp. Wolfe, 1964, p. N30, pl. 12, fig. 3, 1964 = *Populus cedrusensis* Wolfe

This specimen does not have the diagnostic features of *Colubrina*. The two primaries are wavering, whereas those of *Colubrina* are firm and the tertiaries are irregularly quadrangular, not strongly cross-percurrent as in *Colubrina*. Comparison with large suites of *P. cedrusensis* Wolfe leaves shows that they are inseparable.

Daphne septentrionalis (Lesquereux) MacGinitie, 1959, pl. 51, figs. 3, 5, 6, and items in synonymy = *Peraphyllum septentrionalis* (Lesquereux) Axelrod

These leaves from the Florissant flora are inseparable from those produced by *P. ramosissimum*.

Ficus sp. Axelrod, 1939, p. 101, pl. 8, fig. 2 = *Populus sonorensis* Axelrod

That this is a poplar is indicated by the subdued, blunt marginal teeth and the primaries that reach well up into the blade. Several leaves of *P. sonorensis* are in the flora.

Mahonia sinuata Axelrod, new name

Mahonia hollickii Dorf. Arnold, Univ. Mich. Mus. Paleo. Contrib., vol. 5(4), p. 61, pl. 2, figs. 3-5, 7, 8 (Succor Cr.).

Mahonia reticulata (MacGinitie) Brown. Axelrod, Univ. Calif. Pub. Geol. Sci., vol. 33, p. 296, pl. 21, fig. 2 only (Fallon), 1956 (not figs. 1, 3, which are M. macginitiei Axelrod).

Chaney and Axelrod, Carnegie Inst. Wash. Pub. 617, p. 176, pl. 33, figs. 1, 4 (Stinking Water), 1959.

Axelrod, Univ. Calif. Pub. Geol. Sci., vol. 51, p. 121, pl. 13, fig. 2 only (Trapper Cr.), 1964 (not fig. 1, which is M. macginitiei Axelrod).

This species is typified by lanceolate to ovate leaflets with pinnate venation and obliquely cuneate bases. The leaflets have 2-4 widely spaced, broad sinuses separated by lobes with sharp spines. Leaflets of M. eutriphylla (Fedde) Muller from northeastern Mexico show a general resemblance to these fossils. This 4-5 m tall shrub was collected by C. H. Muller in the oak-maple forest of Coahuila, and Hinton secured specimens in the pine woods of Mexico, D.F.

Populus alexanderi Dorf (emend)

Populus alexanderi Dorf, Carnegie Inst. Wash. Pub. 412, p. 75, pl. 7, fig. 3 (Sonoma flora), fig. 11 (Alturas flora), 1930.

Chaney, Carnegie Inst. Wash. Pub. 476, p. 215, pl. 6, figs. 1, 5; pl. 7, fig. 2, 1938 (Deschutes flora).

Axelrod, Carnegie Inst. Wash. Pub. 553, p. 281, pl. 48, fig. 4, 1944 (Alturas flora).

Axelrod, Univ. Calif. Pub. Geol. Sci., vol. 34, p. 128, pl. 19, figs. 1-11, 1958 (Verdi flora).

Dorf (1930) figured specimens from 5 localities that he identified as P. alexanderi, but these represent several taxa as listed below. One of the cotypes (pl. 7, fig. 2) is a typical leaf of P. eotremuloides Knowlton, the other (pl.

7, fig. 3) is here designated the type (syntype) of the species *P. alexanderi*. The species is thus restricted to ovate leaves similar to those produced by *P. trichocarpa* populations in the coastal lowlands from San Francisco Bay southward. These correspond to the type material of *P. trichocarpa* var. *trichocarpa* Torrey and Gray, collected from the Santa Clara River valley near Ventura, coastal southern California. The modern species commonly referred to *P. trichocarpa* has larger leaves that are broadly ovate-lanceolate, often subcordate, and have attenuated apices. That species, which occurs in the mountains of California and ranges northward to Alaska and eastward into the Rocky Mountains, represents *P. hastata* Dode.

P. alexanderi is foreshadowed in the fossil record by *P. emersoni* Condit in the San Pablo flora (Condit, 1938) and it is also in the Carson Pass flora of early middle Miocene age. By contrast, leaves similar to *P. hastata* occur as *P. eotremuloides* Knowlton in the Miocene of the Columbia Plateau region, as in the Succor Creek, Blue Mountains, and Payette floras.

As now restricted, only two of the specimens figured by Dorf (see above) represent *P. alexanderi*. The remainder are assigned as follows:

P. alexanderi Dorf, pl. 6, fig. 9 (plesiotype) =
P. prefremontii Dorf (St. Marys)
P. alexanderi Dorf, pl. 6, fig. 10 (plesiotype) =
P. bonhamii Axelrod (Alturas)
P. alexanderi Dorf, pl. 7, fig. 1 (plesiotype) =
P. prefremontii Dorf (Lafayette Dam)
P. alexanderi Dorf, pl. 7, fig. 2 (cotype) =
P. eotremuloides Knowlton (Wildcat)

Populus lindgreni Knowlton

Populus lindgreni Knowlton, U.S. Geol. Surv. 18th Ann. Rept., pt. 3, p. 725, pl. 100, fig. 3, 1898.

Chaney and Axelrod, Carnegie Inst. Wash. Pub. 617, p. 151, pl. 17, figs. 1-3, 1959.

Wolfe, U.S. Geol. Surv. Prof. Paper 454-N, p. N17, pl. 1, fig. 12, 1964.

Wolfe (1966; Wolfe and Tanai, 1980) placed P. lindgreni Knowlton in synonymy under P. kenaiana Wolfe (see P. voyana, below). However, the type of P. lindgreni from the Payette flora near Montour, Idaho, and similar leaves from the Mascall and Stinking Water (Chaney and Axelrod, 1959) and Fingerrock floras (Wolfe, 1964) differ from P. kenaiana in having a very thick petiole, and the teeth are regularly larger, crenately serrate, and incurved, not sharply acute as in kenaiana. P. kenaiana as figured by Wolfe is allied to P. tremuloides, but lindgreni is more nearly related to P. heterophylla Linnaeus. P. kenaiana as figured by Wolfe (1966) and Wolfe and Tanai (1980) differs in no fundamental way from P. voyana Chaney and Axelrod (1959). Both differ from the usual variation of P. tremuloides in that the fossils have relatively larger teeth, not small crenate ones, or entire leaves, like those produced by aspen. In addition. P. voyana-kenaiana leaves are regularly larger.

Populus subwashoensis Axelrod

Populus subwashoensis Axelrod, Univ. Calif. Pub. Geol. Sci., vol. 33, p. 284, pl. 6, figs. 1-4; pl. 13, figs. 3, 4, 1956.

Axelrod, Univ. Calif. Pub. Geol. Sci., vol. 34, p. 128, pl. 22, figs. 1-4, 1958b.

Populus pliotremuloides Axelrod. Chaney, Carnegie Inst. Wash. Pub. 476, p. 214, pl. 6, fig. 4; pl. 7, figs. 1c, 1d, 1938.

Brown, Wash. Acad. Sci. Jour., vol. 30, p. 226, figs. 20-22, 1949.

This small-leaved species is allied to P. tremula Linnaeus of Eurasia. Therefore, submerging it into P. washoensis Brown (Wolfe, 1964, p. N18) is not acceptable, for that species has consistently large leaves like those of P. grandidentata Michaux.

Populus voyana Chaney and Axelrod

Populus voyana Chaney and Axelrod, 1959, p. 152, pl. 18, figs. 1, 3 4.

Populus kenaiana Wolfe, 1966, p. B12, pl. 3, fig. 1 (not synonymy).

Populus kenaiana Wolfe and Tanai, 1980, p. 35, pl. 11, fig. 9; pl. 12, fig. 1 (not synonymy, except above citation).

Wolfe (1966) included three unrelated poplars under P. kenaiana. The species described initially as Vitis crenata Heer (Flora Fossilis Arctica, p. 36, pl. 8, fig. 1, 1871) was selected as the type of P. kenaiana by Wolfe; the changed epithet involves name priority. Whereas V. crenata, which is the type of P. kenaiana, is a large leaf with relatively large, crenately rounded teeth, those of P. kenaiana figured by Wolfe are smaller and the margin is finely and sharply serrate. The former is allied to P. tomentosa Carriere of northern China, the latter to P. tremuloides Michaux.

Wolfe (1966) and Wolfe and Tanai (1980) also included P. lindgreni Knowlton from the Payette, Mascall, and Stinking Water floras in synonymy under P. kenaiana Wolfe. However, P. lindgreni differs from kenaiana in having a thick petiole and crenate teeth; in most respects it is allied to P. heterophylla Linnaeus, the swamp cottonwood of the eastern United States. One Stinking Water specimen (Chaney & Axelrod, 1959, pl. 17, fig. 1) and some in the Fingerrock flora (Wolfe, 1964) appear to be leaves from sprout shoots.

The relations can be summarized as follows:

- Populus kenaiana Wolfe, type specimen (Vitis crenata Heer) only, is allied to P. tomentosa of China.
- P. voyana Chaney & Axelrod, including figured specimens of P. kenaiana by Wolfe (1966) and Wolfe and Tanai (1980), is allied to P. tremuloides.
- P. lindgreni Knowlton, 1898; Chaney & Axelrod, 1959 and Wolfe (1964) is allied to P. heterophylla.

Robinia lesquereuxi (Etts.) MacGinitie. MacGinitie, 1962, p. 114, pl. 3, fig. 7; pl. 4, fig. 3 = *Robinia kilgoreana*, new sp.

The Kilgore species differs from the Florissant locust (MacGinitie, 1953, p. 127, pl. 34, figs. 1, 2, 4; pl. 46, fig. 2) in its consistently larger leaflets, and these have a more oval outline as compared with the ovate-elliptic leaflets of *R. lesquereuxi* from Florissant. *R. kilgoreana* shows relationship to *R. pseudoacacia* Linnaeus, whereas *R. lesquereuxi* seems more nearly allied to *R. neomexicana* Gray.

Symphoricarpos wassukana Axelrod, 1956, p. 312, pl. 9, fig. 3 only.

This specimen is a legume leaflet that cannot be separated from *Amorpha oblongifolia*, which is also present in Aldrich Station flora.

LITERATURE CITED

ABRAMS, L.

1968 Illustrated Flora of the Pacific States. Stanford University Press. 4 vols.

ADAM, D. P.

1967 Late Pleistocene and Recent palynology in the central Sierra Nevada. In E. J. Cushing and H. E. Wright, Jr. (eds.), Quaternary Paleoecology, p. 275-301. Yale University Press.

1973 Early Pleistocene (?) pollen spectra from near Lake Tahoe, California. U.S. Geol. Surv. Jour. Research 1:691-693.

AHRENDT, L. W. A.

1961 *Berberis* and *Mahonia*: A taxonomic revision. Linnean Soc. London Jour., vol. 57. 410 p.

ARNOLD, C. A.

1936 Some fossil species of *Mahonia* from the Tertiary of eastern and southeastern Oregon. Univ. Michigan Mus. Paleo. Contrib. 5(4):55-66.

AXELROD, D. I.

1939 A Miocene flora from the western border of the Mohave Desert. Carnegie Inst. Wash. Pub. 516. 128 p.

1940 Late Tertiary floras of the Great Basin and border areas. Bull. Torrey Bot. Club 67:477-487.

1944 The Sonoma flora. Carnegie Inst. Wash. Pub. 553: 167-206.

1950a A Sonoma florule from Napa, California. Carnegie Inst. Wash. Pub. 590:23-71.

1950b The Anaverde flora of southern California. Carnegie Inst. Wash. Pub. 590:119-158.

1956 Mio-Pliocene floras from west-central Nevada. Univ. Calif. Pub. Geol. Sci. 33:1-316.

1957a Paleoclimate as a measure of isostasy. Amer. Jour. Sci. 255:690-696.

1957b Age-curve analysis of angiosperm floras. Jour. Paleontol. 31:273-280.

1958a Evolution of the Madro-Tertiary Geoflora. Bot. Review 24:433-509.

1958b The Pliocene Verdi flora of western Nevada. Univ. Calif. Pub. Geol. Sci. 34:91-160.

1962 A Pliocene _Sequoiadendron_ forest from western Nevada. Univ. Calif. Pub. Geol. Sci. 39:195-268.

1964 The Miocene Trapper Creek flora of southern Idaho. Univ. Calif. Pub. Geol. Sci. 51:1-119.

1965 A method for determining the altitudes of Tertiary floras. The Paleobotanist 14:144-171.

1966a The Eocene Copper Basin flora of northeastern Nevada. Univ. Calif. Pub. Geol. Sci. 59:1-119.

1966b The Pleistocene Soboba flora of southern California. Univ. Calif. Pub. Geol. Sci. 60:1-109.

1968 Tertiary floras and topographic history of the Snake River basin, Idaho. Geol. Soc. Amer. Bull. 79:713-734.

1976a History of the conifer forests, California and Nevada. Univ. Calif. Pub. Botany 70:1-62.

1976b Evolution of the Santa Lucia fir (_Abies bracteata_) ecosystem. Missouri Bot. Garden Ann. 63:24-41.

1979 Age and origin of Sonoran Desert vegetation. Calif. Acad. Sci. Occas. Papers 132:1-74.

1980 Contributions to the Neogene paleobotany of central California. Univ. Calif. Pub. Geol. Sci. 121:1-212.

1981 Altitudes of Tertiary forests estimated from paleotemperature. In Proc. Sympos. on Qinghai-Xizang (Tibet) Plateau: Geol. and Ecol. Studies of Qinghai-Xizang Plateau, vol. 1:131-137. Science Press, Beijing, and Gordon and Breach, New York.

AXELROD, D. I., and H. P. BAILEY

1976 Tertiary vegetation, climate and altitude of the Rio Grande depression, New Mexico-Colorado. Paleobiology 2:235-254.

AXELROD, D. I., and W. S. TING

1961 Early Pleistocene floras from the Chagoopa surface, southern Sierra Nevada. Univ. Calif. Pub. Geol. Sci. 39:119-194.

BAILEY, H. P.

1960 A method of determining the warmth and temperateness of climate. Geografiska Annaler 42:1-16.

1964 Toward a unified concept of the temperate climate. Geog. Review 54:516-545.

BANDY, O. L., and J. C. INGLE, Jr.

1970 Neogene planktonic events and radiometric scale, California. Geol. Soc. Amer. Spec. Paper 124: 131-172.

BARRON, J. A.

1973 Late Miocene-early Pliocene paleotemperatures for California from marine diatom evidence. Paleogeog., Paleoclimatol., Paleoecology 14: 277-291.

BARROWS, K. J.

1971 Geology of the southern Desatoya Mountains, Churchill and Lander Counties, Nevada. Ph.D. thesis, University of California, Los Angeles. 349 p.

BECKER, H. F.

1961 Oligocene plants from the upper Ruby basin, southwestern Montana. Geol. Soc. Amer. Mem. 82. 127 p.

1969 Fossil plants of the Tertiary Beaverhead basins in southwestern Montana. Palaeontographica 127 (Abt. B). 142 p.

BERRY, E. W.

1934 Miocene plants from Idaho. U.S. Geol. Surv. Prof. Paper 185-E:97-123.

BONHAM, H. F.

1969 Geology and mineral deposits of Washoe and Storey Counties, Nevada. Nevada Bur. Mines Bull. 70: 1-170.

BROWN, R. W.

1937 Additions to some fossil floras of the western United States. U.S. Geol. Surv. Prof. Paper 186-J:163-206.

1950 Cretaceous plants from southwestern Colorado. U.S. Geol. Surv. Prof. Paper 221-D:45-66.

BURKE, D. G., and E. H. McKEE

1979 Mid-Cenozoic volcano-tectonic troughs in central Nevada. Geol. Soc. Amer. Bull. 90:181-184.

BURROWS, C. J.

1980 Long-distance dispersal of plant macrofossils. New Zealand Jour. Botany 18:321-322.

CHANEY, R. W.

1938 The Deschutes flora of eastern Oregon. Carnegie Inst. Wash. Pub. 476:185-216.

1944 The Dalles flora. Carnegie Inst. Wash. Pub. 553: 285-321.

1959 Miocene floras of the Columbia Plateau. I. Composition and interpretation. Carnegie Inst. Wash. Pub. 617:1-134.

CHANEY, R. W., and D. I. AXELROD

1959 Miocene floras of the Columbia Plateau. II. Systematic considerations. Carnegie Inst. Wash. Pub. 617:135-229.

CLEMENTS, F. E.

1920 Plant indicators: the relation of plant communities to process and practice. Carnegie Inst. Wash. Pub. 290:1-388.

CONDIT, C.
1938 The San Pablo flora of west central California. Carnegie Inst. Wash. Pub. 476:217-268.
1944a The Remington Hill flora. Carnegie Inst. Wash. Pub. 553:21-55.
1944b The Table Mountain flora. Carnegie Inst. Wash. Pub. 553:57-90.

COOPER, W. S.
1922 The broad-sclerophyll vegetation of California: an ecological study of the chaparral and its related communities. Carnegie Inst. Wash. Pub. 319:1-124.

DORF, E.
1930 Pliocene floras of California. Carnegie Inst. Wash. Pub. 412:1-112.
1936 A late Tertiary flora from southwestern Idaho. Carnegie Inst. Wash. Pub. 476:73-124.

DURHAM, J. W.
1950 Cenozoic marine climates of the Pacific coast. Geol. Soc. Amer. Bull. 61:1243-1264.

DURRELL, C.
1966 Tertiary and Quaternary geology of the northern Sierra Nevada. Calif. Div. Mines and Geol. Bull. 190:185-197.

EDWARDS, S. W.
1983 Cenozoic history of Port Orford and Alaskan *Chamaecyparis* cedars. Ph.D. thesis, University of California, Berkeley. 271 p.

EVERNDEN, J. F., and G. T. JAMES
1964 Potassium-argon dates and the Tertiary floras of North America. Amer. Jour. Sci. 262:945-974.

FEDDE, F.
1901 Versuch einer Monographie der Gattung *Mahonia*. Bot. Jahrb. 31:30-133.

FRAKES, L. A., and E. M. KEMP

1972 Influence of continental positions on early Tertiary climates. Nature 240:97-100.

1973 Paleogene continental positions and evolution of climate. In D. H. Tarling and S. K. Runcorn (eds.), Implications of Continental Drift to the Earth Sciences, vol. 1:539-559. Academic Press, New York and London.

FROELICH, H. A., D. H. McNABB, and F. GAWEDA

1982 Average annual precipitation in southwest Oregon, 1960-1980. Oregon State Univ. Ext. Serv. Misc. Pub. 8220.

GRAHAM, A.

1965 The Succor Creek and Trout Creek Miocene floras of southeastern Oregon. Kent State Univ. Bull., Research Ser. 9:1-147.

GRIFFIN, J. R., and W. B. CRITCHFIELD

1972 The distribution of forest trees in California. USDA Forest Serv. Research Paper PSW-82/1972. 118 p. (Reprinted, with supplement, 1976.)

HALL, C. A.

1964 Shallow-water climates and molluscan provinces. Ecology 45:226-234.

HEDGPETH, J. W.

1957 Marine biogeography. Geol. Soc. Amer. Mem. 67: 359-382.

HEER, O.

1871 Flora fossilis Alaskana: flora fossilis Arctica, 2(2):1-41. Also issued in Kgl. svenska vetensk. akad. Handl. 8(4):1-41, 1869.

HOLLICK, A.

1936 The Tertiary floras of Alaska. U.S. Geol. Surv. Prof. Paper 182. 185 p.

HU, H. H., and R. W. CHANEY

1940 A Miocene flora from Shantung Province, China. Carnegie Inst. Wash. Pub. 507. 147 p.

INGLE, J. C., Jr.
1977 Summary of Neogene planktic foraminiferal biofacies, biostratigraphy, and paleooceanography of the marginal North Pacific Ocean [abstract]. In Proc. 1st Internat. Congr. on Pacific Neogene Stratigraphy (Tokyo, 1976), p. 177-182. Kaiyo Shuppan Co., Tokyo.

JEPSON, W. L.
1910 The Silva of California. Univ. Calif. Mem. 2. 480 p.

KELLER, G., and J. A. BARRON
1983 Paleoceanographic implications of Miocene deep-sea hiatuses. Geol. Soc. Amer. Bull. 94:590-613.

KENNETT, J. P.
1980 Paleooceanographic and biogeographic evolution of the Southern Ocean during the Cenozoic, and Cenozoic microfossil datums. Paleogeog., Paleoclimatol., Paleoecology 31:123-152.
1981 Marine tephrochronology. In C. Emiliani (ed.), The Sea, vol. 7. The Oceanic Lithosphere, p. 1373-1436. Wiley, New York.

KENNETT, J. P., A. R. McBIRNEY, and R. C. THUNELL
1977 Episodes of Cenozoic volcanism in the circum-Pacific region. Jour. Volcanology and Geothermal Research 2:145-163.

KENNETT, J. P., and R. C. THUNELL
1975 Global increase in Quaternary explosive volcanism. Science 187:497-503.

KLUCKING, E. P.
1959 The fossil Betulaceae of western North America. M.A. thesis, University of California, Berkeley. 166 p.

KNOWLTON, F. H.
1898 The fossil plants of the Payette Formation. U.S. Geol. Surv. 18th Ann. Rept., pt. 3, p. 721-744.
1902 Fossil flora of the John Day basin, Oregon. U.S. Geol. Surv. Bull. 204. 153 p.

1923 Fossil plants from the Tertiary lake beds of south-central Colorado. U.S. Geol. Surv. Prof. Paper 131-G:183-197.

LaMOTTE, R. S.

1936 The upper Cedarville flora of northwestern Nevada and adjacent California. Carnegie Inst. Wash. Pub. 455:47-142.

LEOPOLD, E. B., and H. D. MacGINITIE

1972 Development and affinities of Tertiary floras in the Rocky Mountains. In A. Graham (ed.), Floristics and Paleofloristics of Asia and Eastern North America, p. 147-200. Elsevier Publ. Co., New York.

LESQUEREUX, L.

1878 Report on the fossil plants of the auriferous gravel deposits of the Sierra Nevada. Mem. Harvard Mus. Comp. Zool. 6(2):1-62.

1887 List of recently identified fossil plants belonging to the United States National Museum, with descriptions of several new species. U.S. Nat. Mus. Proc. 10:21-46.

MacGINITIE, H. D.

1933 The Trout Creek flora of southeastern Oregon. Carnegie Inst. Wash. Pub. 416:21-68.

1953 Fossil plants of the Florissant Beds, Colorado. Carnegie Inst. Wash. Pub. 599. 188 p.

1969 The Green River flora of northwestern Colorado and northeastern Utah. Univ. Calif. Pub. Geol. Sci. 83:1-203.

1974 An early Middle Eocene flora from the Yellowstone-Absaroka volcanic province, northwestern Wind River basin, Wyoming. Univ. Calif. Pub. Geol. Sci. 108:1-103.

MAHER, L. J., Jr.

1964 Ephedra (and Sarcobatus) pollen in sediments of the Great Lakes region. Ecology 45:391-395.

MAPEL, W. J., and W. J. HAIL, Jr.
1959 Tertiary geology of the Goose Creek district, Cassia County, Idaho; Box Elder County, Utah; and Elko County, Nevada. U.S. Geol. Surv. Bull. 1055-H:217-254.

MASON, H. L.
1927 Fossil records of some west American conifers. Carnegie Inst. Wash. Pub. 346:139-158.

MASTROGIUSEPPE, R. J.
1980 A study of _Pinus balfouriana_ Grev. and Balf. (Pinaceae). Systematic Botany 5:86-104.

NOBLE, D. C.
1972 Some observations of the Cenozoic volcano-tectonic evolution of the Great Basin, western United States. Earth and Planet Sci. Letters 17:142-150.

OWENS, J. N., and M. MOLDER
1974 Cone initiation and development before dormancy in yellow cedar (_Chamaecyparis nootkatensis_). Canadian Jour. Bot. 52:2075-2084.

RENNEY, K. M.
1972 The Miocene Temblor flora of west central California. M.S. thesis, University California, Davis. 106 p.

RIEHLE, J. R., E. H. McKEE, and R. C. SPEED
1972 Tertiary volcanic center, west-central Nevada. Geol. Soc. Amer. Bull. 83:1383-1396.

ROUANE, P.
1973 Etude comparée de la répartition des ramifications au cours de l'ontogènese de quelques cupressaées. Trav. Lab. Forest Toulouse, tome 1, vol. 9, art. 3.

SAVIN, S. M., R. G. DOUGLAS, and F. STEHLI
1975 Tertiary marine paleotemperatures. Geol. Soc. Amer. Bull. 86:1499-1510.

SCHOLL, D. W., M. S. MARLOW, N. S. MacLEOD, and E. C. BUFFINGTON

1976 Episodic Aleutian Ridge igneous activity: implications of Miocene and younger submarine volcanism west of Buldir Island. Geol. Soc. Amer. Bull. 87:547-554.

SMILEY, C. J.

1963 The Ellensburg flora of Washington. Univ. Calif. Pub. Geol. Sci. 35:159-276.

SMILEY, C. J., and C. REMBER

1981 Paleoecology of the Miocene Clarkia Lake (northern Idaho) and its environs. In J. Gray et al. (eds.), Communities of the Past, p. 551-590. Hutchinson Ross Publ. Co., Stroudsburg, Pa.

SMILEY, F. J.

1921 A report on the boreal flora of the Sierra Nevada. Univ. Calif. Pub. Botany 9:1-423.

SMITH, H. V.

1941 A Miocene flora from Thorn Creek, Idaho. Amer. Midland Natur. 25:472-522.

SUDWORTH, G. B.

1934 Poplars, principal tree willows, and walnuts of the Rocky Mountain region. USDA Tech. Bull. 420. 112 p.

1967 Forest trees of the Pacific Slope. Dover, New York. 455 p.

TAGGART, R. E., and A. T. CROSS

1980 Vegetation change in the Miocene Succor Creek flora of Oregon and Idaho: a case study in paleosuccession. In D. L. Dilcher and T. N. Taylor (eds.), Biostratigraphy of Fossil Plants, p. 185-210. Dowden, Hutchinson, and Ross, Stroudsburg, Pa.

VOGT, P. R.

1979 Global magmatic episodes: new evidence and implications for the steady-state mid-oceanic ridge. Geology 7:93-98.

WEBBER, I. E.
1930 Woods from the Ricardo Pliocene of Last Chance Gulch, California. Carnegie Inst. Wash. Pub. 412: 113-134.

WILLDEN, R., and R. C. SPEED
1974 Geology and mineral deposits of Churchill County, Nevada. Nevada Bur. Mines and Geol. Bull. 83:1-95.

WOLFE, J. A.
1964 Miocene floras from Fingerrock Wash, southwestern Nevada. U.S. Geol. Surv. Prof. Paper 454-N. 36 p.
1966 Tertiary plants from the Cook Inlet region, Alaska. U.S. Geol. Surv. Prof. Paper 398-B. 32 p.
1969 Neogene floristic and vegetational history of the Pacific Northwest. Madrono 20:83-110.
1972 An interpretation of Alaskan Tertiary floras. In A. Graham (ed.), Floristics and Paleofloristics of Asia and Eastern North America, p. 201-233. Elsevier Publ. Co., New York.

WOLFE, J. A., and D. M. HOPKINS
1967 Climatic changes recorded by Tertiary land floras in northwestern North America. In K. Hatai, Tertiary Correlation and Climatic Changes in the Pacific, p. 67-76. 11th Pacific Sci. Congr., Japan.

WOLFE, J. A., and T. TANAI
1980 The Miocene Seldovia Point flora from the Kenai Group, Alaska. U.S. Geol. Surv. Prof. Paper 1105. 52 p.

PLATES

PLATE 1

Fossil localities in the Middlegate basin, Nevada

Fig. 1. View of Middlegate locality looking southeast to the Eastgate locality in the light-colored shales at the base of the Eastgate Hills. The Middlegate flora occurs in siliceous shales 10-15 m below the top of the Middlegate Formation.

Fig. 2. View of the Eastgate locality in the middle foreground. The Middlegate site is 8 km (5 mi.) distant in the light-colored shales at the right. The contact with the Monarch Mill Formation is visible on the near slope, 10 m above the plant horizon.

PLATE 2

Sclerophyll forest allied to that in the fossil floras

Fig. 1. Sclerophyll forest in the Santa Lucia Mountains, elevation near 800 m (2,600 ft.). Species of *Arbutus*, *Lithocarpus*, and *Quercus* dominate the community. The branch at the right edge of the view is *Q. shrevei*, allied to *Q. shrevoides* in the fossil floras. Diverse riparian taxa in the nearby valleys, distributed in *Acer*, *Amelanchier*, *Platanus*, *Populus*, *Prunus*, *Ribes*, and *Salix*, have allied species in the fossil floras.

Fig. 2. Sclerophyll forest on the east slope of Cone Peak, Santa Lucia Mountains, elevation near 1,370 m (4,500 ft.). The tall spire-like trees in the valley are *Abies bracteata*, associated with *Arbutus*, *Lithocarpus*, and *Quercus*, and with *Pinus coulteri* and *P. lambertiana* scattered here at the upper margin of sclerophyll forest. Nearby chaparral-covered slopes have species of *Ceanothus*, *Cercocarpus*, *Heteromeles*, and *Rhamnus* allied to taxa in the fossil floras.

PLATE 3

Sierra redwood groves with allied taxa in the fossil floras

Fig. 1. Tule River Grove near Belknap Creek, Tulare County, California, altitude 1,500 m (5,000 ft.). Other conifers in view are *Abies concolor*, *Pinus lambertiana*, *P. ponderosa*, and *Calocedrus decurrens*. Their associates with allies in the fossil floras include species of *Acer*, *Aesculus*, *Ceanothus*, *Cercocarpus*, and *Quercus*, as well as species of *Platanus*, *Populus*, *Ribes*, and *Salix* along the river margin. Others that are higher up in the valley include *Abies magnifica*, *Pinus contorta*, *Chrysolepis sempervirens*, and *Populus tremuloides*.

The densely-foliaged trees in the foreground are *Quercus chrysolepis*, whose fossil ally is abundantly represented in the fossil floras. *Cercocarpus*, also commonly represented among the fossil remains, forms dense thickets on nearby exposed slopes, some close to the river.

Fig. 2. North Grove, *Sequoiadendron* forest, Placer County, California, altitude 1,500 m (5,000 ft). Associates in the forest include *Abies concolor*, *Calocedrus decurrens*, *Pinus lambertiana*, *P. ponderosa*, *Pseudotsuga menziesii*, and *Quercus kelloggii*. Shrubs in the area and nearby with equivalents in the Middlegate basin floras include species of *Acer*, *Amelanchier*, *Chrysolepis*, *Lithocarpus*, *Ribes*, and *Salix*. Sclerophyll forest is in the nearby area at slightly lower levels on warmer, south-facing slopes.

MIDDLEGATE FOSSILS

Plate 4

Figs. 1-5. *Abies laticarpus* MacGinitie. Hypotypes 6400-04.

Figs. 6-7. *Picea sonomensis* Axelrod. Hypotypes 6421-22.

Figs. 8-10. *Picea lahontense* MacGinitie. Hypotypes 6412-14.

Figs. 11-13. *Tsuga mertensoides* Axelrod. Hypotypes 6443-45.

Figs. 14-15. *Abies scherri* Axelrod. Paratypes 5491-92.

Fig. 16. *Pseudotsuga sonomensis* Dorf. Hypotype 6441.

Fig. 17. *Pinus sturgisii* Cockerell. Hypotype 6432.

Figs. 18-20. *Sequoiadendron chaneyi* Axelrod. Hypotypes 6446-48.

Fig. 21. *Chamaecyparis cordillerae* Edwards and Schorn. Hypotype 6452.

Fig. 22. *Juniperus nevadensis* Axelrod. Hypotype 6456.

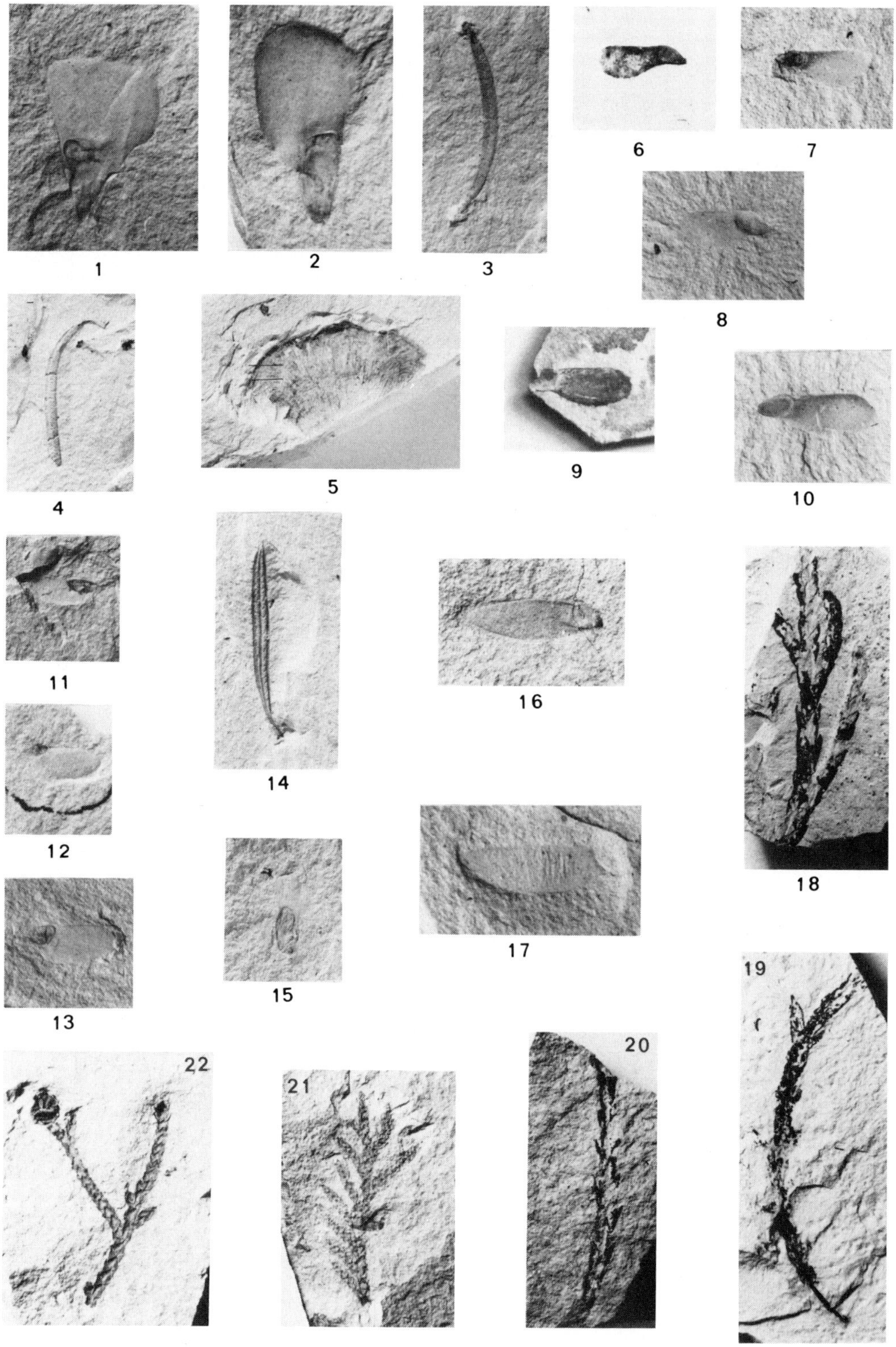
1
2
3
6
7
8
4
5
9
10
11
14
16
18
12
13
15
17
19
20
21
22

MIDDLEGATE FOSSILS

Plate 5

Figs. 1, 3. *Populus payettensis* (Knowlton) Axelrod. Hypotypes 6475-76.

Fig. 2. *Populus cedrusensis* Wolfe. Hypotype 6470.

Figs. 4, 6. *Populus bonhamii* Axelrod. Paratypes 6463-64.

Fig. 5. *Populus bonhamii* Axelrod. Holotype 6462.

1

2

3

4

5

6

MIDDLEGATE FOSSILS

Plate 6

Fig. 1. *Populus bonhamii* Axelrod. Paratype 6465.

Fig. 2. *Salix venosiuscula* Smith. Hypotype 6497.

Figs. 3-4. *Salix storeyana* Axelrod. Hypotypes 6486-87.

Fig. 5. *Populus eotremuloides* Knowlton. Hypotype 6471.

Figs. 6-7. *Salix pelviga* Wolfe. Hypotypes 6480-81.

1
2
3
4
5
6
7

MIDDLEGATE FOSSILS

Plate 7

Fig. 1. *Alnus largei* (Knowlton) Wolfe. Hypotype 6501.

Figs. 2-4. *Betula vera* Brown. Hypotypes 6504-06.

Fig. 5. *Alnus harneyana* Chaney and Axelrod. Hypotype 6500.

1

2

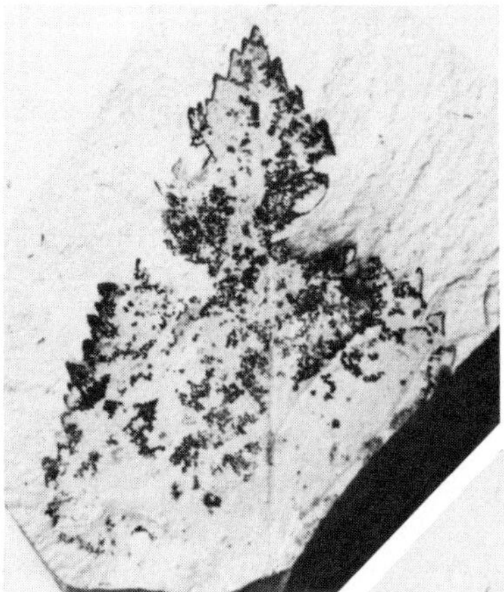

3

4

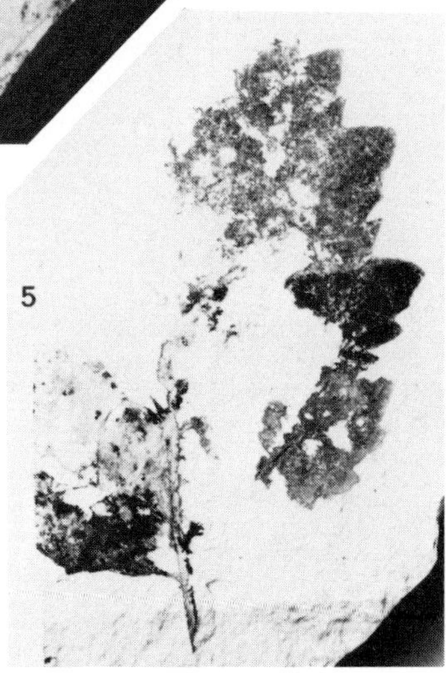

5

MIDDLEGATE FOSSILS

Plate 8

Figs. 1-7. *Quercus hannibali* Dorf. Hypotypes 6537-44.

Figs. 8-10. *Lithocarpus nevadensis* Axelrod. Hypotypes 6522-24.

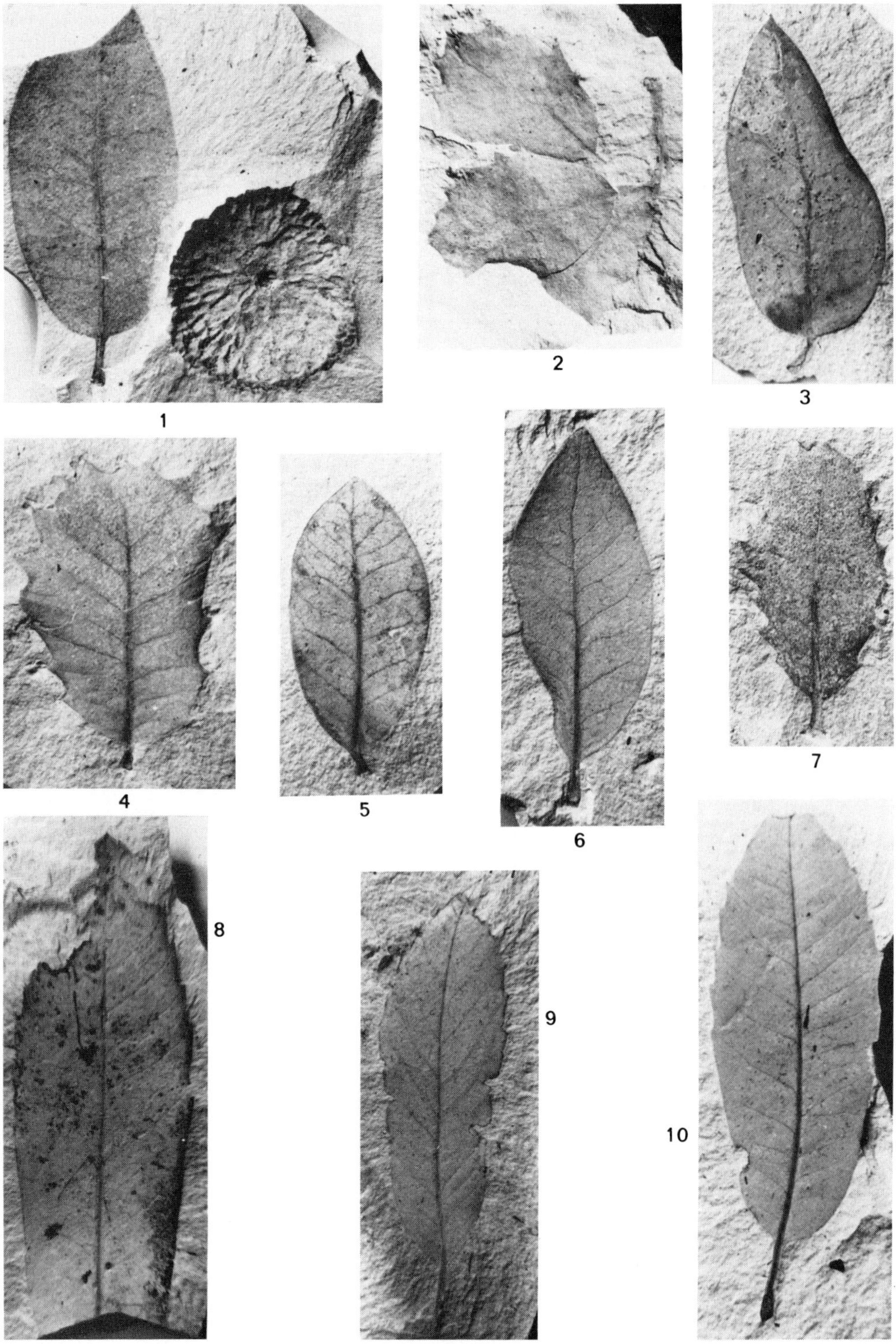
1
2
3
4
5
6
7
8
9
10

MIDDLEGATE FOSSILS

Plate 9

Figs. 1-3. *Chrysolepis convexa* (Lesq.) Axelrod. Hypotypes 6511-13.

Figs. 4-5. *Juglans nevadensis* Berry. Hypotypes (counterparts) 6499a-b.

Fig. 6. *Quercus shrevoides* Axelrod. Hypotype 6570.

Figs. 7-10. *Quercus shrevoides* Axelrod. Paratypes 6571-74.

1
2
3
5
6
7
8
9
10

MIDDLEGATE FOSSILS

Plate 10

Figs. 1-3. *Chrysolepis sonomensis* Axelrod. Hypotypes 6519-21.

Figs. 4-6. *Quercus simulata* Knowlton. Hypotypes 6562-64.

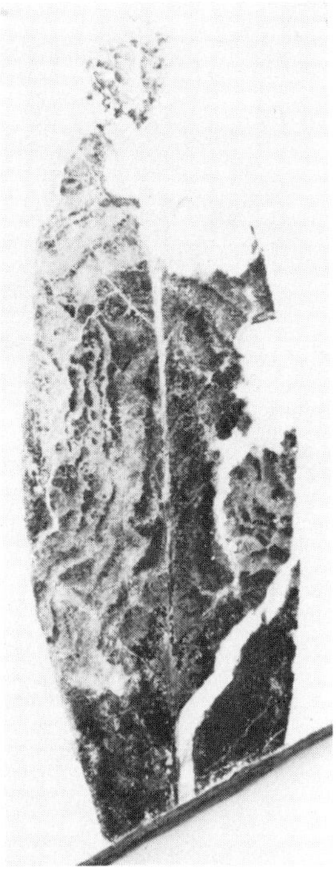

1

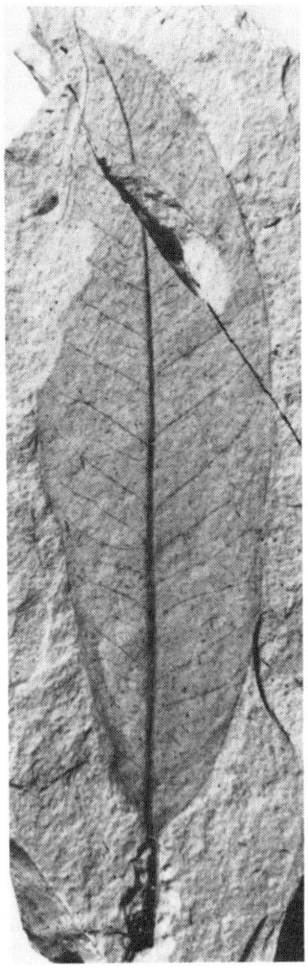

2

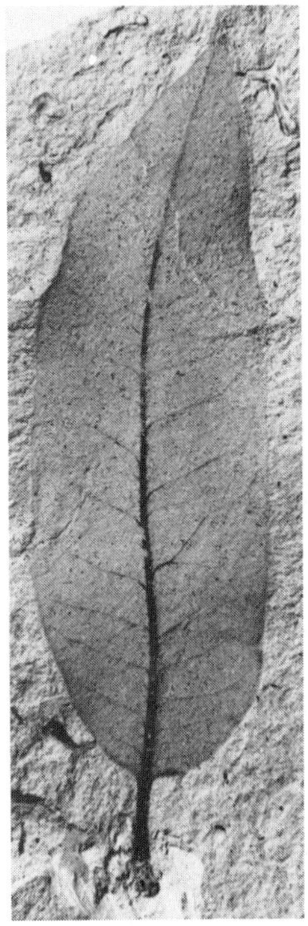

3

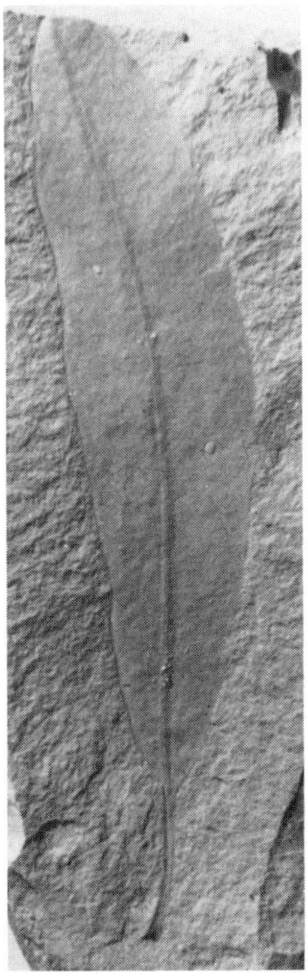

4

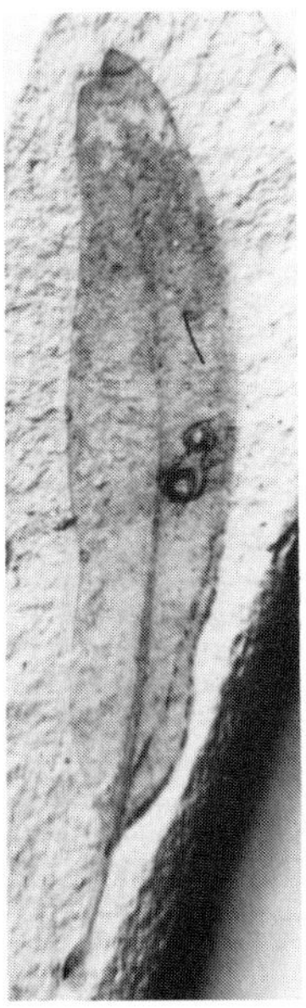

5

6

MIDDLEGATE FOSSILS

Plate 11

Fig. 1. *Mahonia simplex* (Newberry) Arnold. Hypotype 6597.

Figs. 2, 4, 9. *Mahonia macginitiei* Axelrod. Hypotypes 6591-93.

Fig. 3. *Hydrangea ovatifolius* Axelrod. Holotype 6600.

Figs. 5, 10. *Mahonia reticulata* (MacG.) Brown. Hypotypes 6594-95.

Figs. 6-7. *Nymphaeites nevadensis* (Knowlton) Brown. Hypotypes 6598-99.

Fig. 8. *Hydrangea ovatifolius* Axelrod. Paratype 6601.

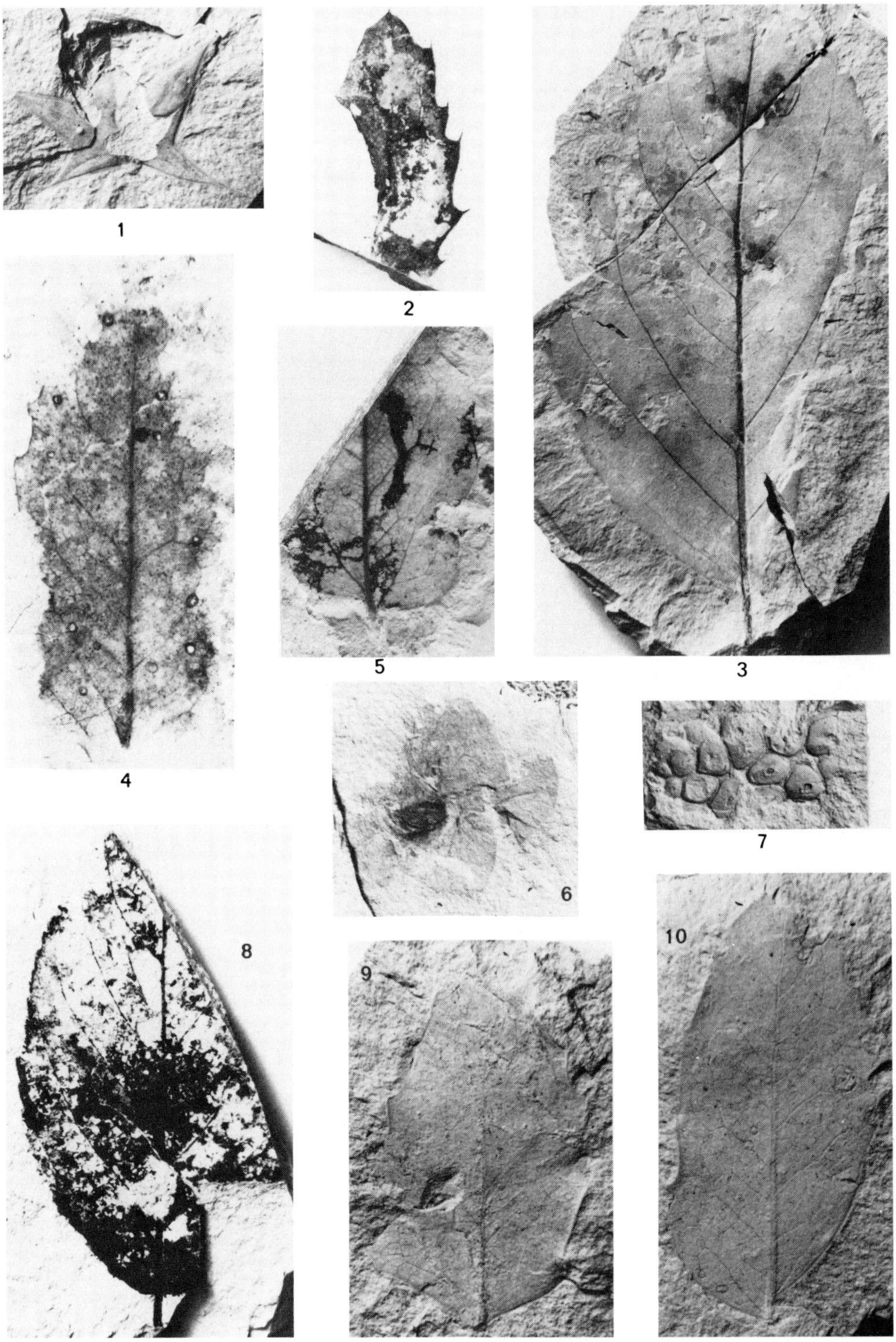
1
2
3
4
5
6
7
8
9
10

MIDDLEGATE FOSSILS

Plate 12

Figs. 1, 4, 9. *Lyonothamnus parvifolius* (Axelrod) Wolfe. Hypotypes 6615-17.

Figs. 2-3. *Crataegus middlegatensis* Axelrod. Hypotypes 6613-14.

Figs. 5-6. *Cercocarpus eastgatensis* Axelrod. Hypotypes 6611-12.

Figs. 7-8, 10. *Heteromeles sonomensis* (Axelrod) Axelrod. Hypotypes 6622-24.

1

2

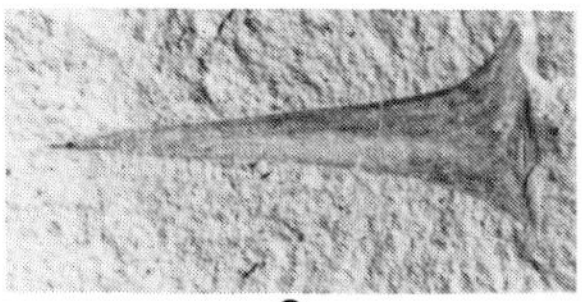

3

4

5

6

7

8

9

10

MIDDLEGATE FOSSILS

Plate 13

Figs. 1-4. *Robinia californica* Axelrod. Hypotypes 6628-31.

Figs. 5-7. *Acer nevadensis* Axelrod. Holotype 6649, paratypes 6650-51.

Figs. 8-9. *Acer oregonianum* Knowlton. Hypotypes 6652-53.

Figs. 10-12. *Cercocarpus antiquus* Lesquereux. Hypotypes 6602-04.

Fig. 13. *Arbutus prexalapensis* Axelrod. Hypotype 6680.

1

2

3

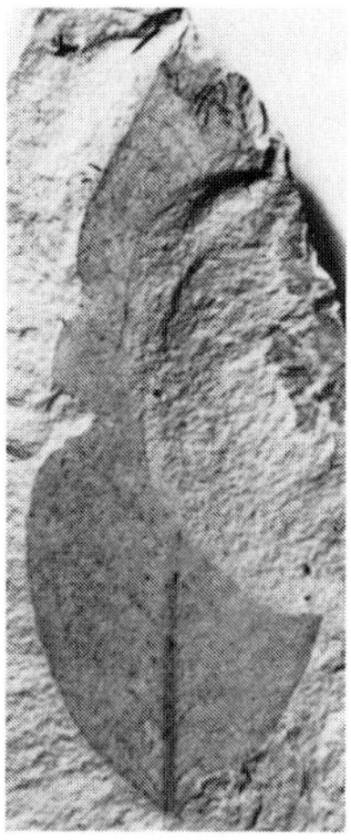
4

5

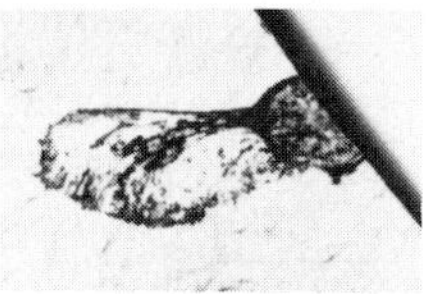
6

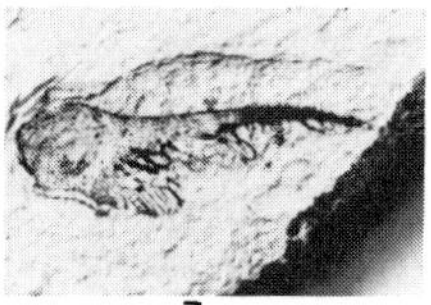
7

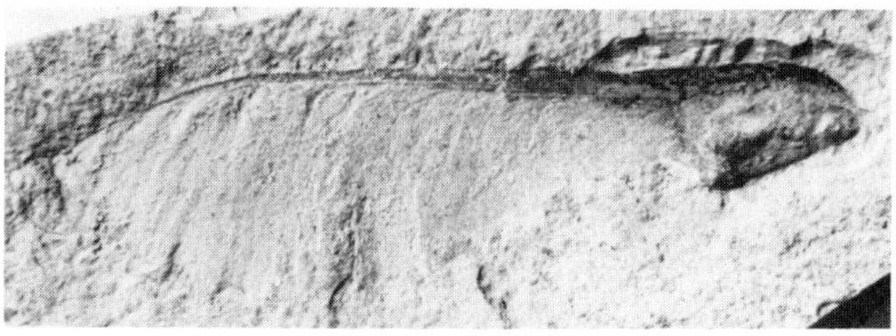
8

9

10

11

12

13

MIDDLEGATE FOSSILS

Plate 14

Figs. 1-6. *Acer tyrrelli* Smiley. Hypotypes 6667-72.

Figs. 7-9. *Acer negundoides* MacGinitie. Hypotypes 6637-39.

Figs. 10-12. *Acer middlegatensis* Axelrod. Hypotypes 6643-45.

1

7

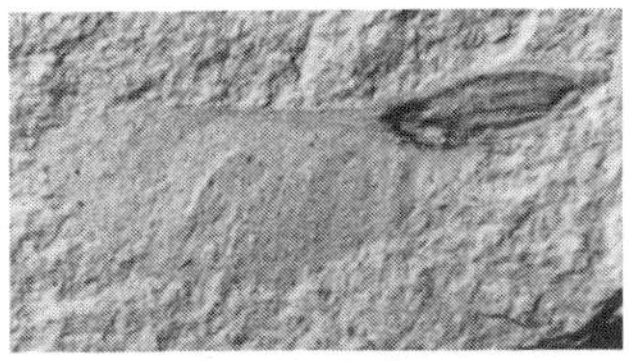
8

9

10

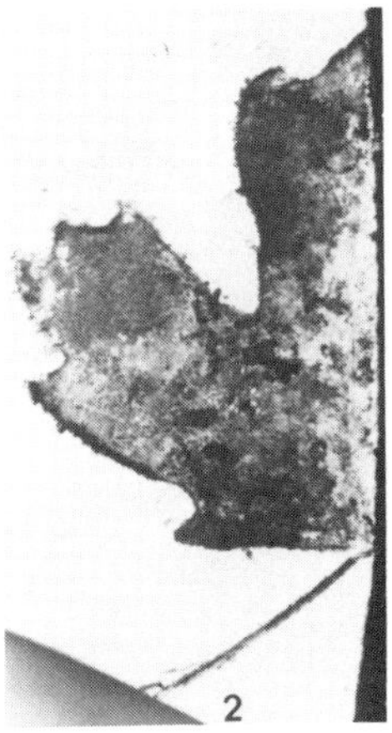
2

4

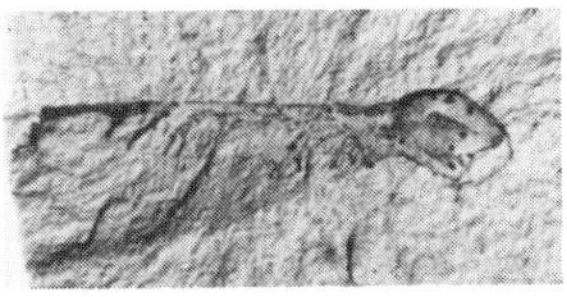
5

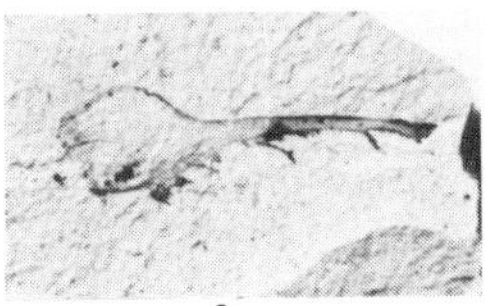
6

3

11

12

MIDDLEGATE FOSSILS

Plate 15

Fig. 1. *Arbutus prexalapensis* Axelrod. Hypotype 6681.

Figs. 2-3. *Gymnocladus dayana* (Knowlton) Chaney and Axelrod. Hypotypes 6627a-b.

Figs. 4-5. *Acer oregonianum* Knowlton. Hypotypes 6654-55.

Figs. 6-8. *Cedrela trainii* Arnold. Hypotypes 6634-36.

Fig. 9. *Diospyros oregonianum* (Lesq.) Chaney and Axelrod. Hypotype 6679.

Figs. 10-11. *Acer scottiae* MacGinitie. Hypotypes 6664-65.

1

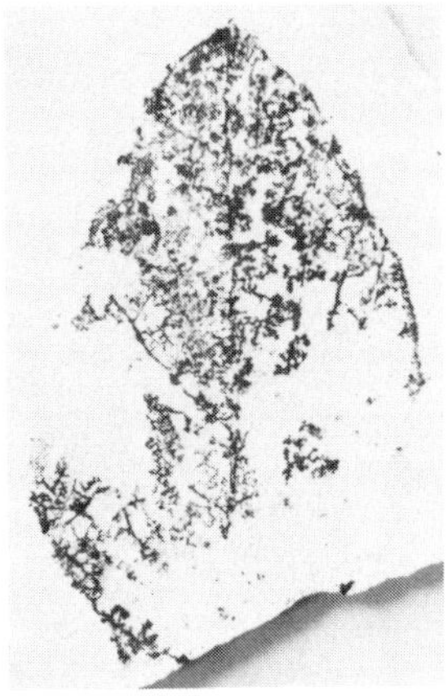

2

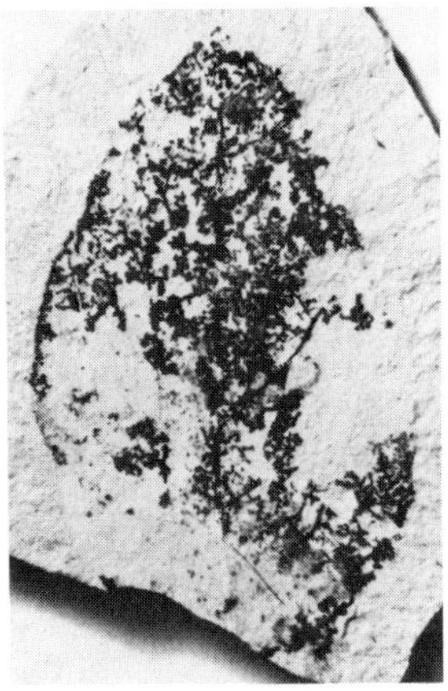

3

4

9

5

6

10

7

11

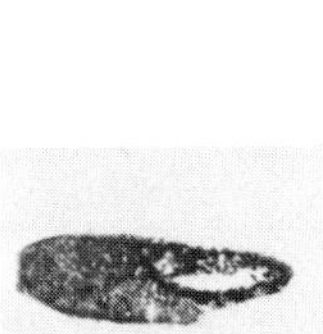

8

EASTGATE FOSSILS

Plate 16

Figs. 1-4. *Abies concoloroides* Brown. Hypotypes 6684-87.

Figs. 5-12. *Abies laticarpus* MacGinitie. Hypotypes 6690-97.

Figs. 13-14. *Pinus balfouroides* Axelrod. Hypotypes 6770-71.

Figs. 15-19. *Pseudotsuga sonomensis* Dorf. Hypotypes 6788-91.

Figs. 20-23. *Tsuga mertensioides* Axelrod. Hypotypes 6804, 6805 (counterpart), 6806-07.

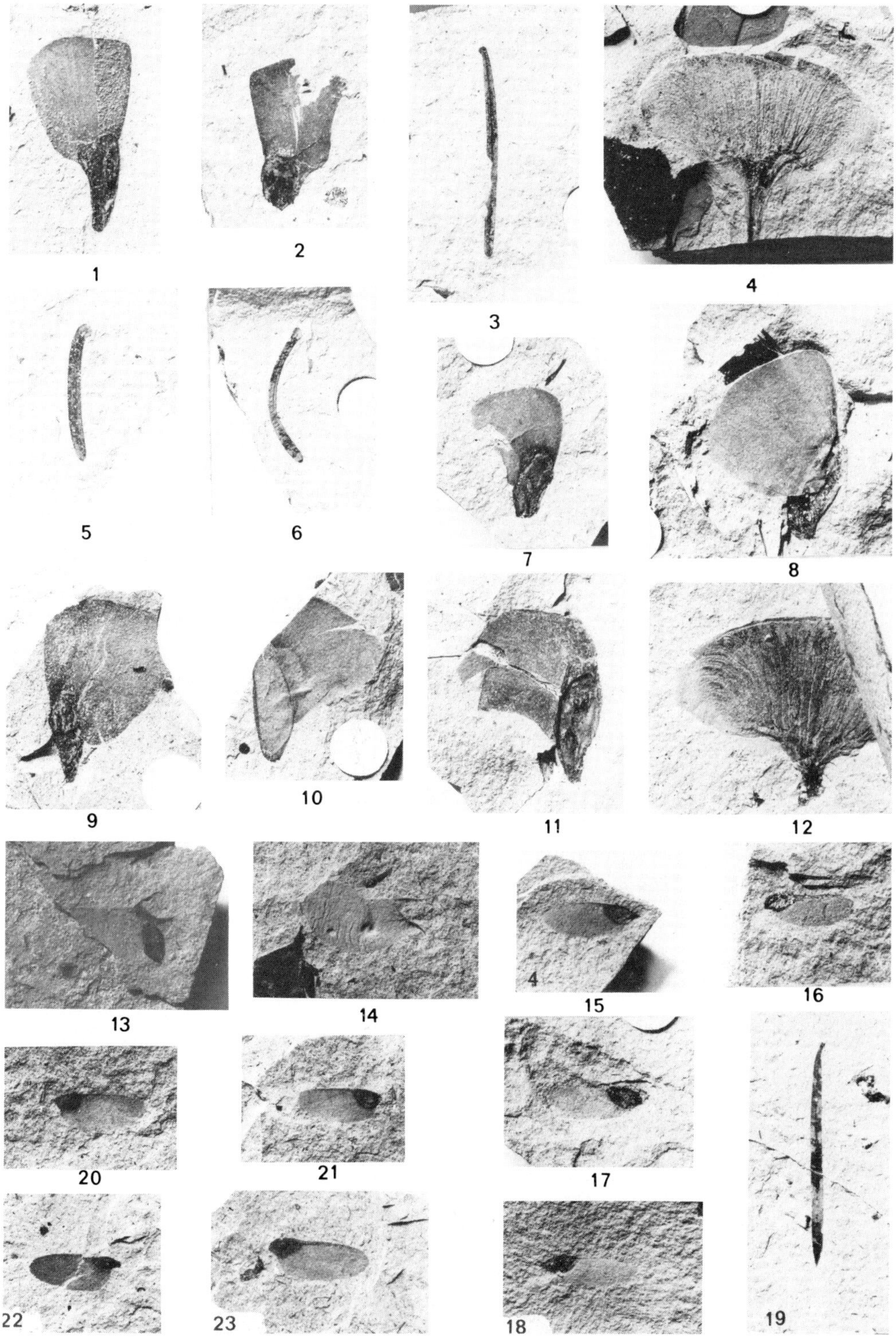
1
2
3
4
5
6
7
8
9
10
11
12
13
14
15
16
20
21
17
22
23
18
19

EASTGATE FOSSILS

Plate 17

Figs. 1-2. *Larix cassiana* Axelrod. Hypotypes 6705-06.

Figs. 3-6. *Picea sonomensis* Axelrod. Hypotypes 6738-41.

Figs. 7-10. *Picea lahontense* MacGinitie. Hypotypes 6721-24.

Figs. 11-14. *Larix nevadensis* Axelrod. Holotype 6708, paratypes 6709-11.

Figs. 15-19. *Pinus alvordensis* Axelrod. Hypotypes 6757-61.

Figs. 21-23. *Chamaecyparis cordillerae* Edwards and Schorn. Hypotypes 6842-44.

Figs. 20, 24. *Juniperus nevadensis* Axelrod. Hypotypes 6850-51.

Figs. 25-26. *Pinus sturgisii* Cockerell. Hypotypes 6777-78.

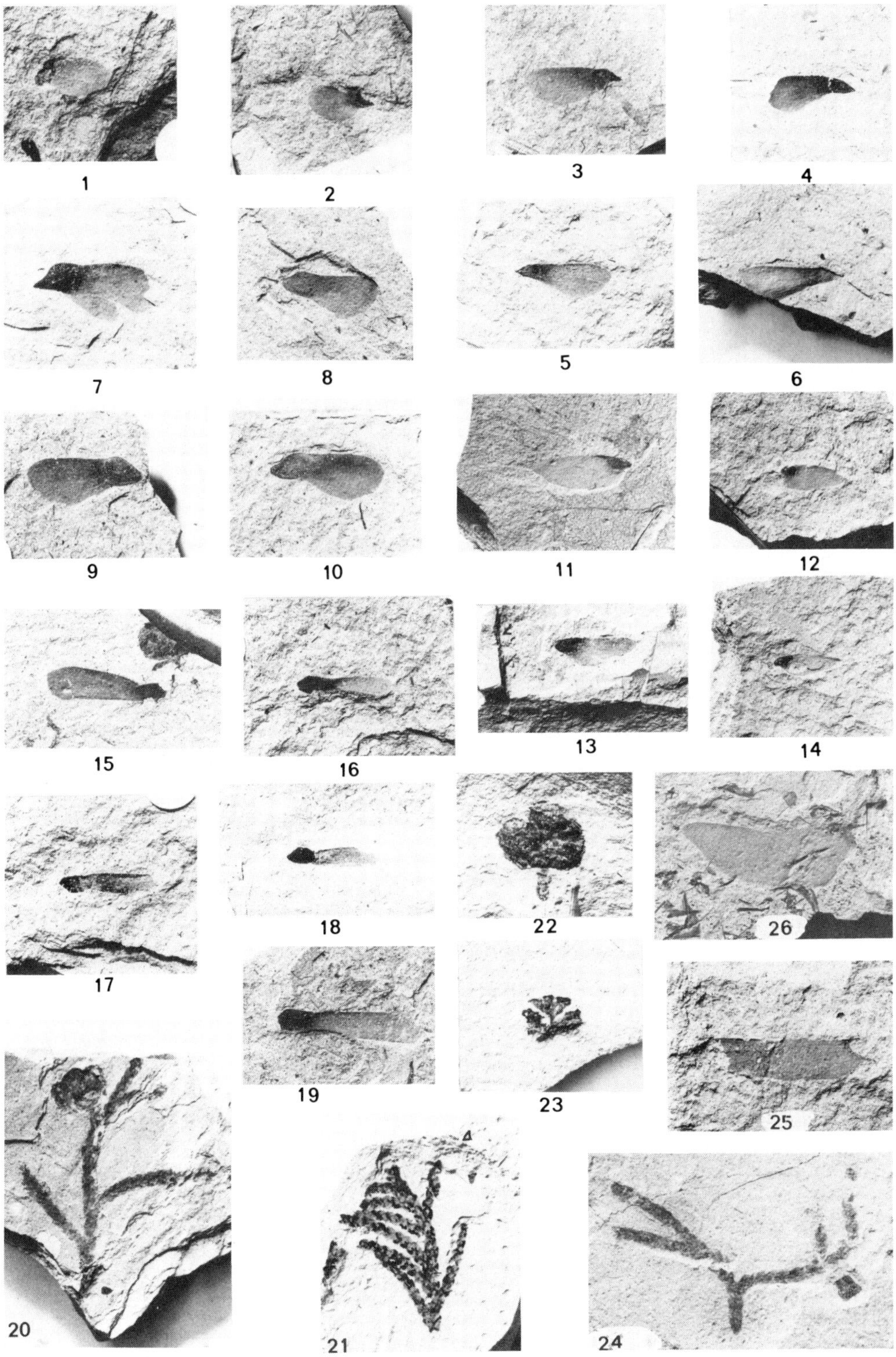
1
2
3
4
7
8
5
6
9
10
11
12
15
16
13
14
17
18
22
26
19
23
25
20
21
24

EASTGATE FOSSILS

Plate 18

Figs. 1-4. *Sequoiadendron chaneyi* Axelrod. Hypotypes 6821-24.

1

2

3

4

EASTGATE FOSSILS

Plate 19

Figs. 1-5. *Sequoiadendron chaneyi* Axelrod. Hypotypes 6825-29.

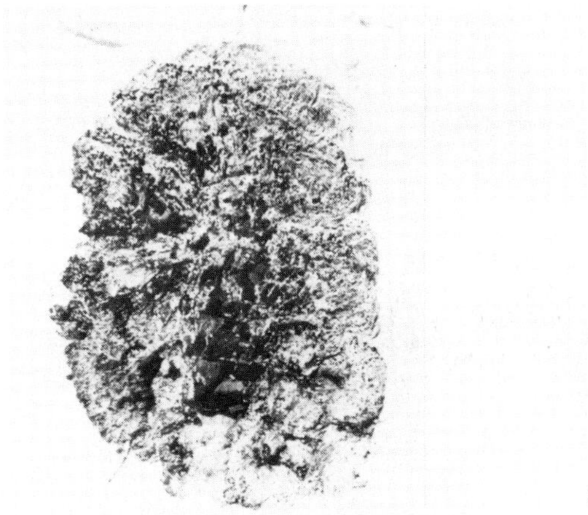

1

2

3

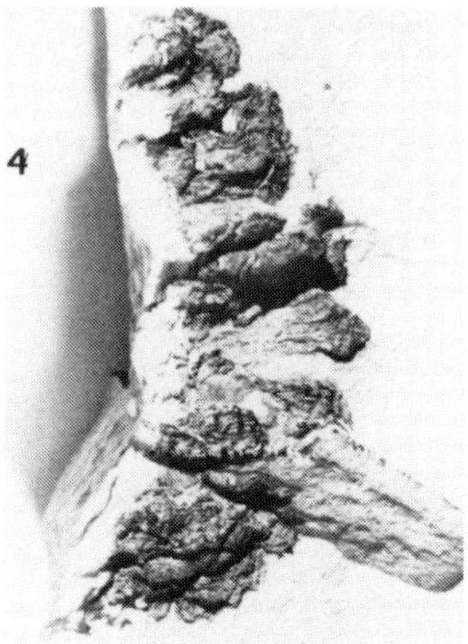

4

5

EASTGATE FOSSILS

Plate 20

Figs. 1, 4, 7. *Populus bonhamii* Axelrod. Hypotypes 6860-62.

Figs. 2, 3, 5. *Populus pliotremuloides* Axelrod. Hypotypes 6876-78.

Fig. 6. *Populus eotremuloides* Knowlton. Hypotype 6871.

1

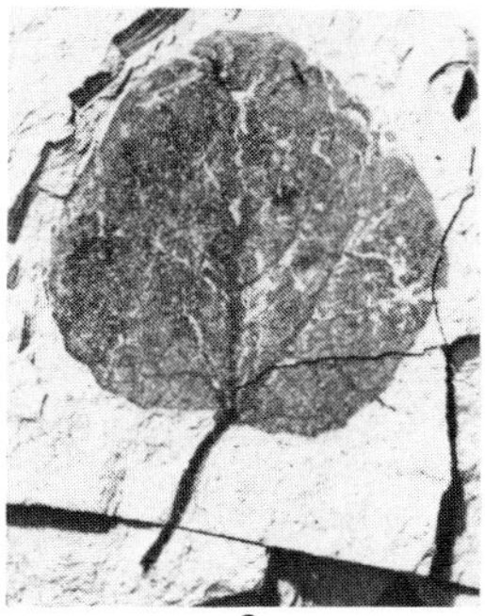

2

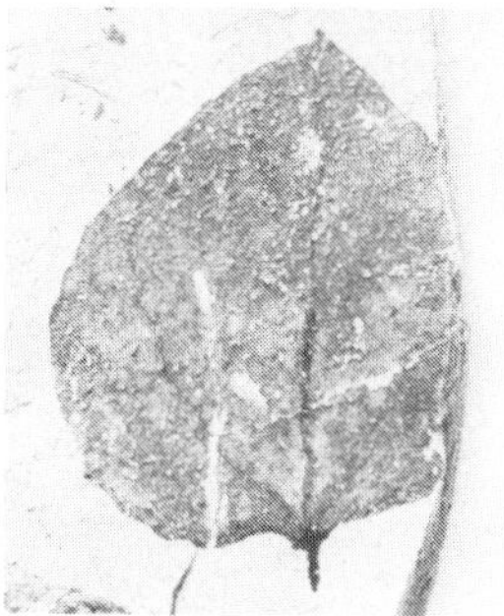

3

4

6

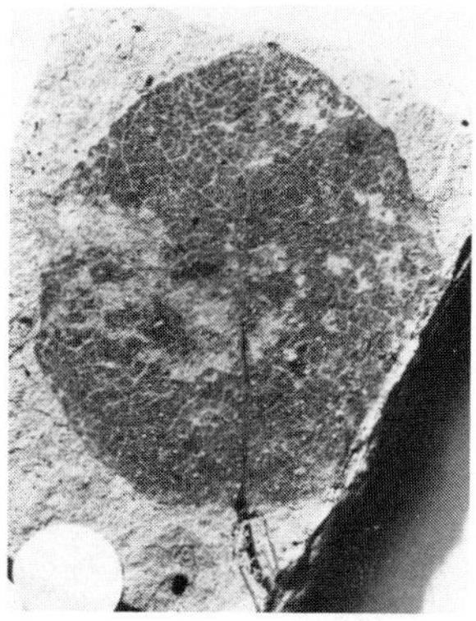

5

7

EASTGATE FOSSILS

Plate 21

Figs. 1-3. *Populus cedrusensis* Wolfe. Hypotypes 6864-66.

Figs. 4-5. *Populus payettensis* (Knowlton) Axelrod. Hypotypes 6873-74.

Fig. 6. *Populus eotremuloides* Knowlton. Hypotype 6872.

1

2

3

4

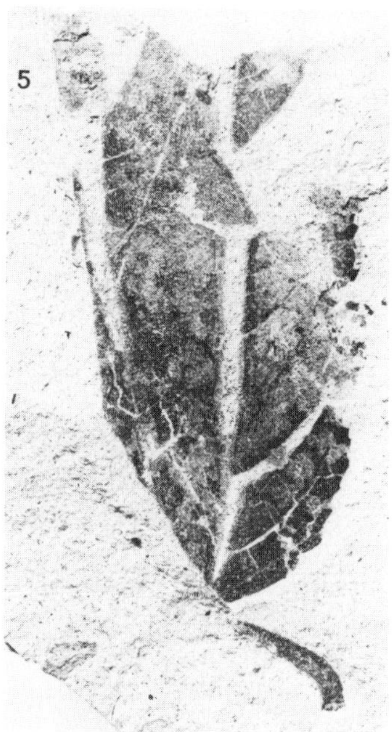

5

6

EASTGATE FOSSILS

Plate 22

Fig. 1. *Typha lesquereuxi* Cockerell. Hypotype 6854.

Figs. 2, 7. *Salix desatoyana* Axelrod. Paratypes 6882-83.

Fig. 3. *Populus payettensis* (Knowlton) Axelrod. Hypotype 6875.

Figs. 4, 8, 12. *Salix pelviga* Wolfe. Hypotypes 6884-86.

Figs. 5, 10. *Ceratophyllum praedemersum* Ashlee. Hypotypes 6977-78.

Fig. 6. *Salix desatoyana* Axelrod. Buffalo Canyon, holotype 6881.

Fig. 9. *Nymphaeites nevadensis* (Knowlton) Brown. Hypotype 7005.

Fig. 11. *Salix venosiuscula* Smith. Hypotype 6896.

1
2
3
4
5
6
7
8
9
10
11
12

EASTGATE FOSSILS

Plate 23

Figs. 1-2. *Chrysolepis convexa* (Lesq.) Axelrod. Hypotypes 6919-20.

Figs. 3, 6. *Betula thor* Knowlton. Hypotypes 6899-6900.

Figs. 4-5. *Chamaecyparis cordillerae* Edwards and Schorn. Winged seeds. Hypotypes 6915-16.

Figs. 7-8. *Salix storeyana* Axelrod. Hypotypes 6891-92.

Fig. 9. *Betula vera* Brown. Hypotype 6917.

1
2
3
4
5
6
7
8
9

EASTGATE FOSSILS

Plate 24

Figs. 1-3, 5-7. *Lithocarpus nevadensis* Axelrod. Hypotypes 6932-34, 69-36-38.

Fig. 4. *Lithocarpus nevadensis* Axelrod. Acorn cup. Hypotype 6935.

1
2
3
4
5
6
7

EASTGATE FOSSILS

Plate 25

Figs. 1-6. *Quercus hannibali* Dorf. Hypotypes 6242-47.

Fig. 7. *Quercus hannibali* Dorf. Acorn cup. Hypotype 6248.

Figs. 8-10. *Quercus shrevoides* Axelrod. Hypotypes 6970-72.

1
2
3
4
5
6
7
8
9
10

EASTGATE FOSSILS

Plate 26

Figs. 1-5. *Mahonia reticulata* (MacGinitie) Brown. Hypotypes 6982-86.

Figs. 6-9. *Quercus simulata* Knowlton. Hypotypes 6961-64.

1
2
3
4
5
6
7
8
9

EASTGATE FOSSILS

Plate 27

Figs. 1-2. *Mahonia reticulata* (MacGinitie) Brown. Hypotypes 6987-88.

Figs. 3-6, 9. *Mahonia simplex* (Newberry) Arnold. Hypotypes 6995-99.

Fig. 7. *Mahonia macginitiei* Axelrod. Hypotype 6981.

Fig. 8. *Lyonothamnus parvifolius* (Axelrod) Wolfe. Hypotype 7055.

EASTGATE FOSSILS

Plate 28

Fig. 1. *Nymphaeites nevadensis* (Knowlton) Brown. Rootstock and rootlets. Hypotype 7006.

Fig. 2. *Amelanchier grayi* Chaney. Hypotype 7011.

Figs. 3, 4. *Nymphaeites nevadensis* (Knowlton) Brown. Leaves. Hypotypes 7007-08.

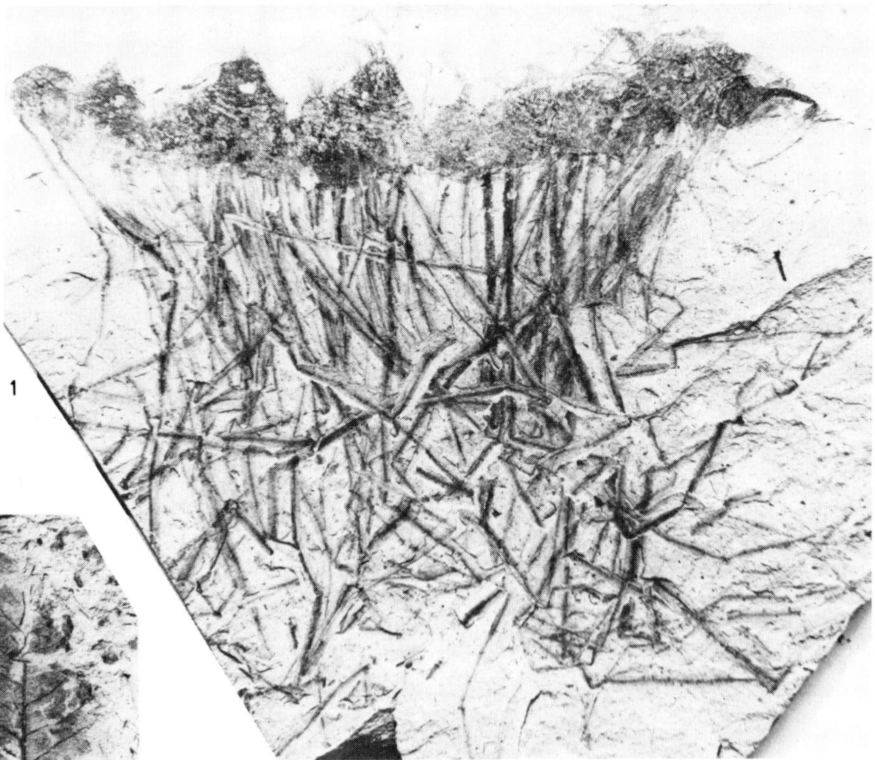

1

2

3

4

EASTGATE FOSSILS

Plate 29

Figs. 1, 3. *Ribes stanfordianum* Dorf. Hypotypes 7009-10.

Figs. 2, 6-7, 10. *Sorbus idahoensis* Axelrod. Hypotypes 7071-74.

Fig. 4. *Sparganium nevadense* Axelrod. Holotype 7137.

Fig. 5. *Sparganium nevadense* Axelrod. Paratype 7138.

Fig. 8. *Lyonothamnus parvifolius* (Axelrod) Wolfe. Hypotype 7056.

Fig. 9. *Nymphaeites nevadensis* (Knowlton) Brown. Hypotype 7140.

Fig. 11. *Quercus screvoides* Axelrod. Hypotype 7139.

1

2

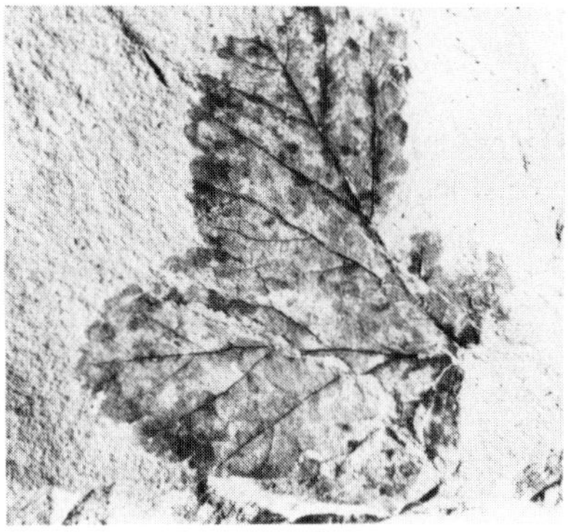

3

4

5

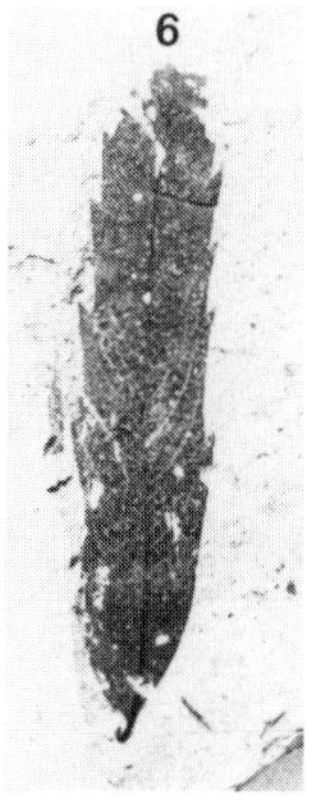

6

7

8

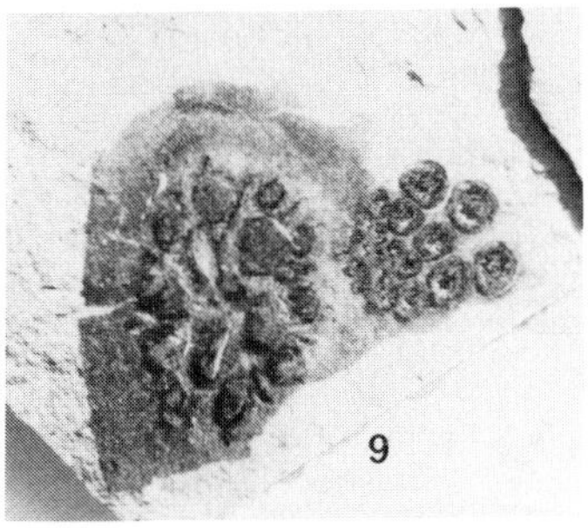

9

10

11

EASTGATE FOSSILS

Plate 30

Figs. 1, 4, 6. *Heteromeles sonomensis* (Axelrod) Axelrod. Hypotypes 7063-65.

Figs. 2-3. *Amelanchier grayi* Chaney. Hypotypes 7012-13.

Fig. 5. *Crataegus newberryi* Cockerell. Hypotype 7054.

Figs. 7-8. *Cercocarpus eastgatensis* Axelrod. Holotypes 7016-16a (counterparts).

Fig. 9. *Prunus chaneyi* Condit. Hypotype 7068.

1
2
3
4
5
6
7
8
9

EASTGATE FOSSILS

Plate 31

Figs. 1-14. *Cercocarpus eastgatensis* Axelrod. Paratypes 7017-30.

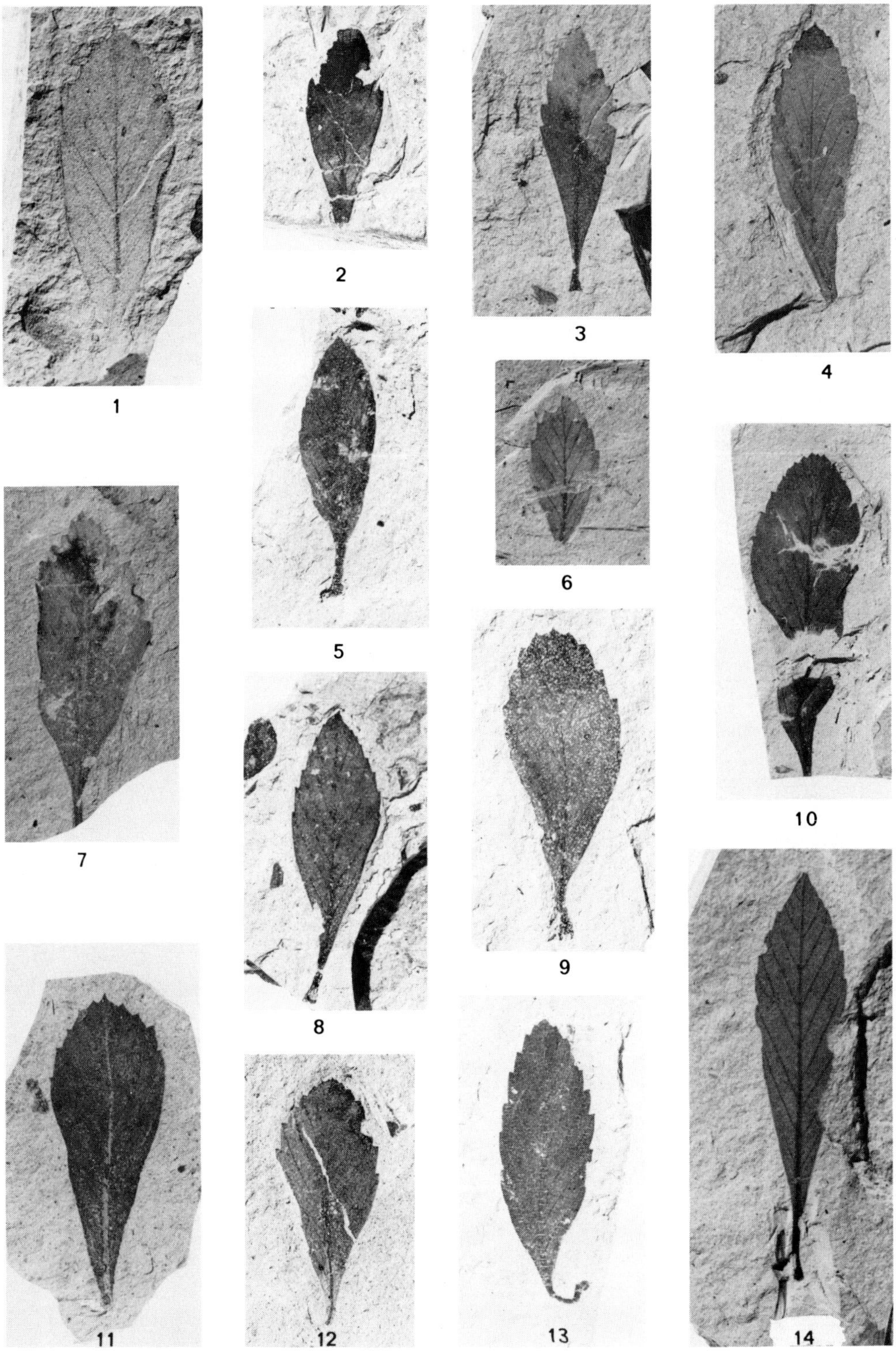
1
2
3
4
5
6
7
8
9
10
11
12
13
14

EASTGATE FOSSILS

Plate 32

Figs. 1-4. *Cercocarpus eastgatensis* Axelrod. Paratypes 7031-32, 7123-24.

Figs. 5-12. *Cercocarpus ovatifolius* Axelrod. Paratypes 7034-41.

Fig. 13. *Cercocarpus ovatifolius* Axelrod. Holotype 7033.

1
2
3
4
5
6
7
8
9
10
11
12
13

EASTGATE FOSSILS

Plate 33

Figs. 1-2. *Robinia californica* Axelrod. Hypotypes 7081, 7067.

Figs. 3, 6-7, 9-12. *Acer tyrrelli* Smiley. Hypotypes 7090, 7092-94, 7084, 7095-96.

Figs. 4-5. *Acer nevadensis* Axelrod. Samara, counterparts. Paratypes 7091-91a.

Fig. 8. *Rhamnus precalifornica* Axelrod. Hypotype 7112.

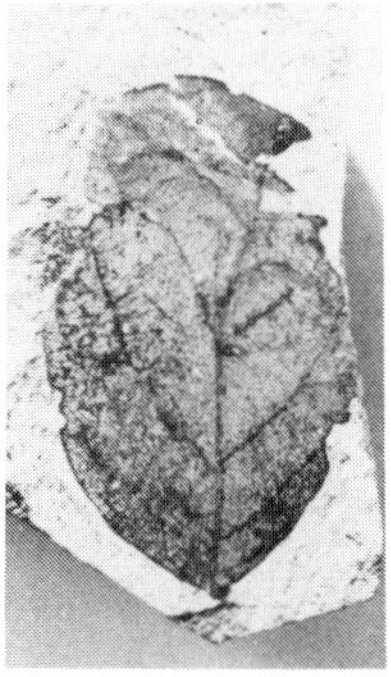
1

2

3

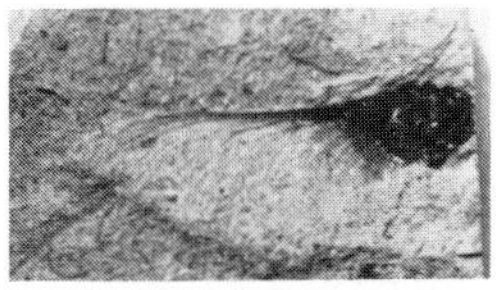
4

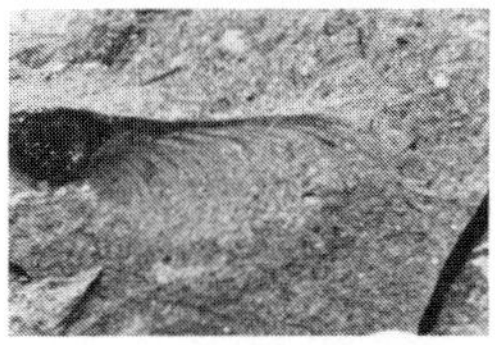
5

6

7

8

9

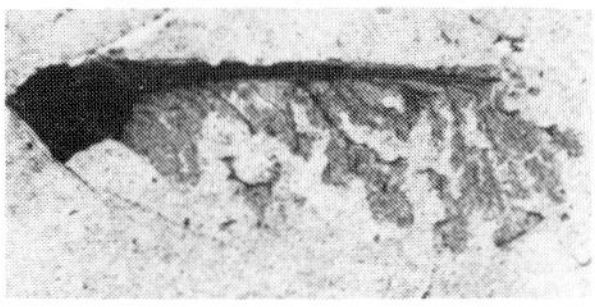
10

11

12

EASTGATE FOSSILS

Plate 34

Figs. 1-2. *Acer oregonianum* Knowlton. Hypotypes 7085-86.

Figs. 3-4. *Acer tyrrelli* Smiley. Hypotypes 7082-83.

Figs. 5-7. *Aesculus preglabra* Condit. Hypotypes 7084, 7106-07.

Fig. 8. *Fraxinus coulteri* Dorf. Hypotype 7121.

Fig. 9. *Eugenia nevadensis* Axelrod. Holotype 7113.

Fig. 10. *Arbutus prexalapensis* Axelrod. Hypotype 7014.

1
2
5
3
9
4
8
6
7
10